Large Carpenter Bees:
A Guide to Species
of *Xylocopa (Neoxylocopa)*
from North and Central America

Large Carpenter Bees:
A Guide to Species
of *Xylocopa (Neoxylocopa)*
from North and Central America

Jonathan Mawdsley

Pineway Press
2017

First Printing: 2017

ISBN-13: 978-1546554417
ISBN-10: 1546554416

Pineway Press
6801 Pineway
University Park, MD, 20782, USA

www.jonathanmawdsley.com

Dedication

This book is dedicated to my good friend and colleague Sam Droege, whose enthusiasm for bees and bee identification is highly contagious, to Corinne Carter for all of her help and encouragement, and to my parents, Ralph and Alice Mawdsley.

Contents

Acknowledgments ix

Abstract - Resumen xi

Preface xiii

Chapter 1: Introduction 1

Chapter 2: Materials and Methods 5

Chapter 3: Diagnosis and Key to Species 7

Chapter 4: Species Accounts 15

References 45

Glossary 49

.

Acknowledgments

I thank my distinguished colleague, good friend, and scientific mentor Terry Erwin for sponsoring my formal affiliation as a Research Associate with the National Museum of Natural History, Smithsonian Institution. My colleagues Seán Brady and Brian Harris are to be thanked for very kindly making the important collections of large carpenter bees in the National Museum of Natural History, Smithsonian Institution, available for study, and for providing me with access to the extensive library of reprints and notebooks that had been assembled by the late Smithsonian entomologist Paul Hurd for his own studies of the genus *Xylocopa*. For contributing additional specimens which were used in this study, I thank the following individuals: David DeRosa, Brent Karner, Miguel Serrano, and Jay Timberlake. I also thank my friends and colleagues in the broader pollinator conservation community for their general encouragement of my studies, particularly Laurie Davies Adams, John Ascher, Scott Hoffman Black, Stephen Buchmann, Gabriela Chavarria, Sam Droege, Paul Goldstein, Jerome Rozen, Dolores Savignano, and Tom Van Arsdall. For assistance with photography and for providing photographic equipment for this project, I thank Alice Mawdsley, and for general advice and suggestions regarding the publication of this book I thank Sam Droege. Corinne Carter and Ralph and Alice Mawdsley provided much valuable encouragement and support throughout this project.

Abstract - Resumen

This guidebook provides materials for the identification of large carpenter bees of the genus *Xylocopa* Latreille, subgenus *Neoxylocopa* Michener, from the southwestern United States, mainland México, and the countries of mainland Central America south to and including Panamá. Keys and diagnoses for the identification of females and males of the ten species of these bees recorded from North and Central America are provided in both English and Spanish. For each species, a full synonymy, brief diagnoses of both sexes, and a summary of the known distribution in North and Central America is provided. Collecting locality data is provided for all museum specimens examined. All species are illustrated in color.

Esta guía presenta materiales para la identificación de grandes abejas carpinteras del género *Xylocopa* Latreille, subgénero *Neoxylocopa* Michener, del suroeste de los Estados Unidos, México continental y los países de América Central a Panamá. Claves y diagnósticos se dan para la identificación de hembras y machos de las diez especies de estas abejas registradas en América del Norte y América Central, tanto en inglés como en español. Para cada especie, hay un sinónimo completo, diagnósticos breves de ambos sexos y un resumen de la distribución conocida en Norte y Centroamérica. Datos de la localidad se dan para todos los especímenes examinados. Todas las especies se ilustran en color.

Preface

This book had its origins on a warm Sunday afternoon in March, 2012, in the beautiful old city of Ponce, Puerto Rico, when I found a large black bee floating in the fountain in the square at the center of town. Intrigued, I fished out the bee, which turned out to be a species of a large and poorly-known group known as "carpenter bees," so-called because the adult bees dig tunnels in dead wood. The discovery of this bee was timely, as I was already thinking at that point that I should broaden my own entomological research program to include studies of at least one group of bees. By the time that I visited Ponce, I had already been studying flower pollination by beetles for over a decade, and had participated actively in several high-profile political advocacy efforts on behalf of pollinators and their conservation needs. However, at that point I had still not yet begun to study what some of my entomological friends and colleagues like to call the "real pollinators," the Apoidea or bees. Finding this single bee specimen was the first step on a path towards the development of a strong interest in the biology and taxonomy of the carpenter bees, with particular attention to the most poorly-known groups species from the world's tropical regions.

This book is about a particular group of carpenter bees and specifically the species in that group that are found from the southwestern United States to southern Panamá. This group or "subgenus" was called "*Neoxylocopa*" by the great bee biologist Charles Michener because most of the species in this group occur in the New World. The genus *Xylocopa*, whose species are generally known as "large carpenter bees," is massive and mega-diverse, with over 700 described species worldwide. Most of these species occur in the tropics and subtropics, and many of these bees are extremely difficult to identify accurately to species. I chose to study this group of bees for two reasons: first, because many of its species are important pollinators in tropical and subtropical ecosystems, and second, because the species in this particular group of bees have been and continue to be extraordinarily difficult to identify. The late Smithsonian entomologist Paul Hurd, who studied the *Neoxylocopa* bees for his entire entomological career, called these the "look-alike" bees. The females of the Central American species of *Neoxylocopa* are black in color with black or brown pubescence (some South American species in this group have reddish-orange or yellow pubescence), while the males are orange or brown with red, orange, or yellow pubescence. The males and females of these bees are extremely dissimilar in appearance, so much so that many earlier scientists concluded that they were actually separate species. Males and females of a species can only be associated by finding them in the same nest, by finding them engaged in reproductive behaviors, or through modern innovative methods such as DNA barcoding.

Although our knowledge of the biology and natural history of these bees is incomplete, available evidence suggests that they are important pollinators of flowering plants in tropical forest systems, particularly trees and shrubs with large, open colorful flowers with copious pollen and/or nectar rewards. I hope that the publication of this guide to the species in this group from North and Central America will stimulate further research on the pollination biology and ecological interactions involving these fascinating bees, which function as important pollinators of crop plants as well as wild plants in tropical forests throughout Central America.

Chapter 1: Introduction

Large carpenter bees, species of the genus *Xylocopa* Latreille (Insecta: Hymenoptera: Apidae), represent one of the larger evolutionary lineages of bees, with over 700 species described to date worldwide and many more species awaiting discovery and description by scientists (Hurd and Moure 1963; Hurd 1978; Michener 2007). Many of the species of the genus *Xylocopa* are important pollinators of agricultural crops, particularly in tropical regions where this genus has its greatest species diversity (Keasar 2010). Crops of economic importance which are pollinated by these bees in the New World tropics or "Neotropics" include passion fruit, *Passiflora edulis* Sims, (Passifloraceae; Dominguez-Gil and McPheron 1992) the Brazil nut tree, *Bertholletia excelsa* Humboldt and Bonpland (Lecythidaceae; Motta Maués 2002), and cotton, *Gossypium* spp. (Malvaceae; Waller et al. 1985). Species of the subgenus *Neoxylocopa* have also been recorded visiting flowers of numerous other flowering plant taxa, including many tropical forest trees as well as agricultural crops such as alfalfa (*Medicago sativa* L., Fabaceae), coffee (*Coffea* sp., Rubiaceae), cowpea (*Vigna unguiculata* (L.) Walpers, Fabaceae), ginger (*Zingiber* sp., Zingiberaceae), pumpkins and squash (*Cucurbita* spp., Cucurbitaceae), sweet pea (*Lathyrus odoratus* L., Fabaceae), and tomato (*Solanum lysopersicum* L., Solanaceae; Hurd 1978).

As indicated by the common name of "carpenter bees," many species of the genus *Xylocopa* excavate their nests in exposed dead wood, including standing dead trees, dead tree limbs, fence posts, exposed rafters, and other wooden structures (Gerling et al. 1989). Damage to human structures is well documented, and sometimes may be severe enough to cause structural failure (Janzen 1966; Hurd 1978). These bees can also become minor pests of ornamental trees in tropical urban areas, particularly when damage from wind, weather, or improper pruning creates injuries to the trees that facilitate nest excavation by the bees (O'Farrill-Nieves and Medina-Gaud 2007). Several species of the subgenus *Neoxylocopa* included in this book are known to cause significant damage to exposed dead wood, including structural timbers (Janzen 1966; Hurd 1978).

Despite the importance of these bees as pollinators and pests of trees and structural timbers, the identification of individual species of *Xylocopa* is often very difficult. Outside of a few well-studied areas such as the United States, South Africa, and Australia, there are few published guides to species of the genus *Xylocopa*. For many areas of the world (including much of tropical America, Asia, and Africa) the only tools available for identifying these bees are the original descriptions of the species which can be found scattered throughout the primary scientific literature. These descriptions, published in many now-obscure journals, were written in a variety of European languages by different authors over the years. The original descriptions are often extremely brief and only mention a few superficial diagnostic characteristics. And few (if any) reliable diagnostic illustrations were provided by the original taxonomists who first studied these species.

To add further confusion, in many tropical areas there are strong similarities between many species of *Xylocopa* in their external coloration (Hurd and Moure 1963; Hurd 1978). Similarities in coloration among the female bees appear to be due (at least in part) to mimetic interactions that may involve multiple species of *Xylocopa* as well as other sympatric bee species and non-stinging insects such as flies (Mawdsley 2015). All females of *Xylocopa* species can sting, a behavior which likely functions as a powerful anti-predator defense. Convergence among multiple species of stinging insects on a common color pattern, a phenomenon known as Müllerian mimicry, can facilitate the general recognition of the species in the mimicry complex by predators and thereby reduce the overall predation pressure on all species involved in the mimicry complex (Wickler 1968).

Males and females of individual *Xylocopa* species can differ significantly in coloration, vestiture, and body proportions, often to such a degree that the different sexes of a single species were often described as separate species by earlier taxonomists (Hurd and Moure

1963; Hurd 1978). In many cases, males and females can only be correctly associated through direct observation of mating or nesting behaviors, by rearing of individuals from the same nest, or by use of modern molecular systematics techniques such as DNA barcoding.

Few subgenera within *Xylocopa* have been as problematic from a taxonomic standpoint as the subgenus *Neoxylocopa* Michener. Males and females of the approximately 50 species in this subgenus are radically different in appearance, making it difficult for taxonomists to accurately associate the males with the females unless the specimens were collected *in copula* or reared from the same nest (Hurd and Moure 1963). Females of many species in this group are similar in appearance, with black integument and black vestiture, while males are also similar in appearance, with brown or orange pigmentation and yellow, red, or orange vestiture (Hurd and Moure 1963; Hurd 1978). The diagnostic characters used by previous authors who studied the species in this group have included the size and density of the punctures on the frons, vertex, mesosoma, and metasoma, the presence or absence of carinae, tubercles, and other raised structures on the frons, the structure of the male genitalia, and the coloration of the wing iridescence, particularly in females (Hurd and Moure 1963; Hurd 1978). Many of these characters have not been studied comprehensively across all the species of the subgenus; for example, Pérez (1901) described *Xylocopa nasica* on the basis of female specimens with a prominent nose-like tubercle on the lower portion of the frons, yet such a structure is also found in the females of many of the other species in this group.

The subgenus *Neoxylocopa* was last treated comprehensively by the late Smithsonian entomologist Paul Hurd (1978), who published an annotated catalogue of species from the New World. Hurd's catalogue also contained considerable new information on the taxonomic status of species in this group, based on his studies of morphological characters of large series of specimens and the available type specimens. Unfortunately, Hurd did not illustrate or provide keys for the identification of many of the species that he studied. This guidebook attempts to remedy these shortcomings of Hurd (1978) by providing new identification materials for the species of this group in one geographic region of the Neotropics (USA, México and Central America south to Panamá). I selected this area because it contains a number of poorly-known species of carpenter bees, and because specimens from this area are particularly well represented in the large collections of specimens of *Xylocopa* species assembled by Hurd at the National Museum of Natural History, Smithsonian Institution, USA. The geographic coverage of this guide extends from the southern United States to the northern boundary of Colombia with Panamá. This area is an important area of endemism for this subgenus of bees: of the ten species included in this work, five are entirely restricted to this part of the New World, and several of these species have even more restricted geographic distributions (see Hurd 1978 and the distribution data and maps presented below).

Hurd and Moure (1963) proposed the subgenus *Megaxylocopa* for the three species *X. fimbriata*, *X. frontalis*, and *X. nautlana*. In this guidebook, I follow Michener (2007), who considered *Megaxylocopa* to be a synonym of the subgenus *Neoxylocopa* Michener. Recent phylogenetic work (Mawdsley 2015) suggests that these two lineages may actually be sister-taxa and thus the three species of *Megaxylocopa* could plausibly be interpreted as representing a separate, monophyletic subgenus, as proposed by Hurd and Moure (1963). The females and males of the three species in what I have termed the "*Megaxylocopa* group" (Mawdsley 2015) are similar in their general appearance and sexual dimorphism to those of the remaining species in subgenus *Neoxylocopa*, and consequently I have elected to include them in this guidebook.

As discussed in more detail in the Materials and Methods section below, there is much that remains to be learned about the biology, behavior, and distribution of species of *Xylocopa* subgenus *Neoxylocopa* which occur in North and Central America. It is hoped that the publication of this brief guide will help to spur further research across this region into all aspects of the natural history of these important pollinating insects.

Chapter 2: Materials and Methods

This guidebook covers the taxonomy and identification of the species of *Xylocopa* subgenus *Neoxylocopa* that are known to occur in the southwestern United States, mainland México, and mainland Central America south to (and including) Panamá. This work is based on my morphological studies of preserved museum specimens of these bees, including the very large collection of carpenter bees which was assembled by the late Paul Hurd at the Department of Entomology, National Museum of Natural History, Smithsonian Institution (NMNH or USNM). This collection includes paratype or syntype specimens of species described by many of the major taxonomists who studied the taxonomy of *Xylocopa*, including T. D. A. Cockerell, J. S. Moure, P. D. Hurd, H. Friese, and J. Pérez. It also includes numerous specimens which were reared by Hurd and colleagues from nests found in the field, an approach which facilitates the accurate association of the males and females of a particular species (Hurd and Moure 1963). I also received specimens of these bees during the course of my studies from a number of private collectors and individuals who are thanked in the Acknowledgments. In addition to the physical specimens, I also examined digital images of type specimens of *Xylocopa* species in the American Museum of Natural History, New York, and the Muséum National d'Histoire Naturelle, Paris.

Morphological nomenclature follows that proposed in the classic study of carpenter bee morphology and taxonomy by Hurd and Moure (1963). A brief glossary is provided at the end of this guidebook which covers those terms that may not be familiar to a general biologist or an entomologist without specialized training in hymenopteran morphology and systematics. Definitions of many of the other standard terms that are used in insect morphology can be found at the Wikipedia page "Glossary of Entomology Terms" < https://en.wikipedia.org/wiki/Glossary_of_entomology_terms>

Our current knowledge of the biology and life history of these bee species is very much incomplete, and further research into all aspects of their biology and behavior is greatly needed. The available flower visitation records and nesting records are summarized here for these species, based on published records in the scientific literature and labels accompanying the specimens that I examined for this study. Hurd (1978) noted that all species in this group for which we have records appear to be polylectic floral visitors, meaning that they visit a broad spectrum of flowering plant species offering suitable nectar and/or pollen rewards. The relatively brief lists of associated flowering plant species and nesting substrates which are included for the species in this present book certainly cannot be considered to be comprehensive. Detailed multi-year field studies have shown that other species of the subgenus *Neoxylocopa* may visit flowers of as many as 285 plant species (Jackson and Woodbury 1976) and utilize as many as 29 different tree species as nesting substrates (Díaz and Sánchez 1998). Additional field studies will be needed in order to identify the full range of floral associates and nesting substrates for the *Neoxylocopa* species included in this guidebook.

Our current knowledge of the distribution of these bees is likewise incomplete. The most up-to-date distribution maps showing the geographic points where specimens of these species have been collected can be found on the "Discover Life" website:
< http://www.discoverlife.org >. These maps summarize the known distribution information for these bees, using data taken from the scientific literature and from specimens in several major museum collections. However, certain portions of tropical America have not been well surveyed for these bees. New distribution records at the state and country levels are therefore very likely, particularly for some of the species that are less well represented in museum collections. For purposes of this guide, I have mapped the known distribution for these species at the state level in México, and at the country level for the other countries covered in the guide. Two species (*X. mexicanorum* and *X. varipuncta*) also occur in portions of the United States, as noted in the key and in the individual accounts for these species.

Chapter 3: Diagnosis and Key to Species

Genus *Xylocopa* Latreille (1802:379)

Type species: *Apis violacea* Linnaeus (1758:578), by subsequent designation of Westwood (1840:56).

Subgenus *Neoxylocopa* Michener (1954:155, 157-158)

Type species: *Apis brasilianorum* Linnaeus (1767:961), by original designation.

Separation from sympatric subgenera of *Xylocopa*: According to Michener (2007), individuals of species of subgenus *Neoxylocopa* can be separated from those of other, sympatric subgenera of *Xylocopa* on the basis of the following characters: **Females:** Integument predominantly or entirely black, with black (rarely brown) vesture; sternites of the metasoma with a strong, longitudinal median carina; frons often with carinae, tubercles and/or projections. **Males:** Strongly dissimilar from the female in overall appearance; integument predominantly or entirely brown, yellow, or orange, with orange, yellow, or brown vesture.

Separación de los otros subgéneros de *Xylocopa*: Según Michener (2007), los individuos de las especies del subgénero *Neoxylocopa* pueden separarse de los de otros subgéneros de *Xylocopa* sobre la base de los siguientes caracteres: **Hembras:** Tegumento predominantemente o enteramente negro, con negra (raramente marrón) pubescencia; esternitos del metasoma con una carina mediana longitudinal fuerte; frons frecuentemente con carinas, tubérculos y / o proyecciones. **Machos:** Fuertemente disímiles de la hembra en apariencia general; integumento predominantemente o enteramente marrón, amarillo o naranja, con vestidor naranja, amarillo o marrón.

Large Carpenter Bees

Key to adults of North and Central American species of *Xylocopa*, subgenus *Neoxylocopa*

1) Integument predominantly or entirely black, with predominantly black vestiture; females... 2
- Integument predominantly or entirely yellowish-brown, reddish-brown, or orange, with orange, yellow or brown vestiture; males... 11

2) Frons with a transverse, raised carina adjacent to or incorporating the ocelli... 3
- Frons lacking such a transverse carina near the ocelli................ 5

3) Frontal carina with lateral projections or "horns". *X. (N.) fimbriata*
- Frontal carina lacking such lateral projections......................... 4

4) Frontal carina interrupted at middle by median ocellus...............
.... *X. (N.) frontalis*
- Frontal carina entire, median ocellus located above carina..........
.... *X. (N.) nautlana*

5). Vestiture of frons and mesosoma entirely reddish-brown; frons mostly flat and lacking a distinct median tubercle between the antennal bases.................................... *X. (N.) ocellaris*
- Vestiture black, with at most brown pubescence at the front of the mesosoma; frons with a more or less distinct median tubercle between the antennal bases.................................... 6

6) Wings hyaline, mostly translucent to transparent, not strongly dark brownish-black, usually with iridescence........................ 7
- Wings very dark brownish-black, not at all strongly transparent, usually with iridescence... 8

7) Wings with brilliant red or red-violet and gold iridescence, México and USA..................................... *X. (N.) varipuncta*
- Wings with indistinct green, blue, and/or violet iridescence, Panamá
.... *X. (N.) columbiensis*

8) Wings with uniformly golden-green iridescence...... *X. (N.) nasica*
- Wings with bluish-green, blue, or violet iridescence.................. 9

9) Wings with strong violet iridescence throughout.....................
.... *X. (N.) mexicanorum*
- Wing iridescence bluish-green with violet tints....................... 10

10) Lateral projection of metatibiae strongly bi-lobed, lobes subequal
 in size, Fig. 10 d................................. *X. (N.) wilmattae*
- Lateral projection of metatibiae with one large lobe and one much
 smaller lobe, Fig. 4 d......................... *X. (N.) gualanensis*

11) Vestiture bright red, lacking elongate lateral setae on metasomal
 tergites 2-3.................................... *X. (N.) columbiensis*
- Vestiture brown, yellowish-brown, or orange, with elongate lateral
 setae on metasomal tergites 2-3................................ 12

12) Metasomal pigmentation uniformly yellowish-brown or reddish-
 brown, lacking distinct transverse bands or patterns of dark
 brown pigmentation.. 13
- Metasomal pigmentation yellowish brown or reddish-brown with
 distinct transverse bands or patterns of dark brown pigmentation... 16

13) Metasomal tergites glabrous medially but with dense orange setae
 laterally... *X. (N.) fimbriata*
- Metasomal tergites more or less uniformly covered with dense
 orange setae... 14

14) Integument dark brownish-yellow; Texas and eastern México
 south to Honduras........................... *X. (N.) mexicanorum*
- Integument bright yellow or yellowish-orange; southwestern United
 States and western México.................... *X. (N.) varipuncta*

16) Metasomal tergites each with a narrow dark brown band along
 apical margin, metasoma predominantly orange-brown or yellowish-brown.. 17

- Metasomal tergites with broad dark brownish-black bands, tergites predominantly dark brownish-black, especially those towards apex of metasoma..…............ 18

17) Mesonotum uniformly dark orange-brown..... *X. (N.) gualanensis*
- Mesonotum dark orange-brown with an inverted bright orange triangle on disc..............................….… *X. (N.) wilmattae*
18) First and second metasomal tergites yellowish-brown at base and with a dark brown band across apical portion, dark brown band larger on second than first tergite, apical tergites entirely dark brown..…... *X. (N.) nasica*
- All metasomal tergites yellowish-brown at base and with dark brown band across apical portion, dark brown bands becoming broader on apical tergites but yellow coloration still present...................................…....................................…..... 19
19) Parameres of male genitalia lacking a ventral tooth in lateral view, Fig. 3 f...…..... *X. (N.) frontalis*
- Parameres of male genitalia with a well-defined ventral tooth visible in lateral view, Fig. 7 g......…....................….. *X. (N.) nautlana*

Clave para adultos de especies Norteamericanas y Centroamericanas de *Xylocopa*, subgénero *Neoxylocopa*

1) Tegumento predominantemente o enteramente negro, con pubescencia predominantemente negra; Hembras................... 2
- Tegumento predominantemente o enteramente marrón amarillento, rojizo-marrón, o anaranjado, con pubescencia anaranjada, amarilla o marrón; Machos.. 11

2) Frons con una carina transversal, levantada adyacente o incurporando los ocelos... 3
- Frons sin una carina transversa cerca de los ocelos................... 5

3) Carina frontal con proyecciones laterales o "cuernos"................
.... *X. (N.) fimbriata*
- Carina frontal sin proyecciones laterals............................... 4

4) Carina frontal interrumpida en el centro por ocelo mediano...........
.... *X. (N.) frontalis*
- Carina frontal entera, ocelo mediano situado sobre la carina...........
.... *X. (N.) nautlana*

5). Pubescencia del frons y de mesosoma enteramente de color marrón rojizo; frons su mayoría planos y carecen de un tubérculo mediana distinta entre los bases de las antenas...............
.... *X. (N.) ocellaris*
- Pubescencia negra, de vez en cuando con la pubescencia marrón en la parte delantera de la mesosoma; frons con un tubérculo más o menos clara la mediana entre las bases de las antenas........ 6

6) Alas hialinas, en su mayor parte translúcidas a transparentes, no muy oscuras de color pardo-negro, por lo general con iridiscencia... 7
- Alas muy oscuras pardo-negro, no muy fuertes transparentes, generalmente con iridiscencia.. 8

7) Alas con brillante rojo o rojo-violeta y oro iridiscencia, México y
    Estados Unidos...............................*X. (N.) varipuncta*
- Alas con verde, azul, y / o violeta iridiscencia, Panamá................
....*X. (N.) columbiensis*

8) Alas con iridiscencia uniformemente dorada-verde... *X. (N.) nasica*
- Alas con iridiscencia azul-verdosa, azul o violeta.................... 9

9) Alas con fuerte iridiscencia violeta............. *X. (N.) mexicanorum*
- Iridiscencia de ala azulada-verde con tonos violetas................. 10

10) Proyección lateral de metatibiae fuertemente en dos lóbulos,
     lóbulos subequales en tamaño, Fig. 10 d...... *X. (N.) wilmattae*
- Proyecto lateral de metatibiae con dos lóbulos, un lóbulo grande y
     un lóbulo más pequeño, Fig. 4 d.............. *X. (N.) gualanensis*

11) Pubescencia rojo brillante, carente de setas laterales alargadas en
     tergitos metasómicos 2-3.................... *X. (N.) columbiensis*
- Pubescencia marrón, marrón amarillento o naranja, con setas later-
     ales alargadas en tergitos metasómicos 2-3.................... 12

12) Pigmentación metasomal uniformemente marrón amarillento o
     marrón rojizo, carente de bandas transversales o patrones de
     pigmentación marrón oscuro.................................... 13
- Pigmentación metasomal de color marrón amarillento o marrón ro-
     jizo con distintas bandas transversales o patrones de pig-
     mentación de color marrón oscuro............................ 15

13) Tergitos metasomal glabro medialmente pero con setas naranjas
     densos lateralmente............................. *X. (N.) fimbriata*
- Tergitos metasomal más o menos uniformemente cubiertos con setas
     naranjas densos... 14

14) Tegumento amarillo parduzco oscuro; Texas y el este de México
     hacia el sur a Honduras.................... *X. (N.) mexicanorum*
- Tegumento amarillo brillante o amarillo-naranja; Al sudoeste de los
     Estados Unidos y el oeste de México......... *X. (N.) varipuncta*

15) Tergitos metasómicos cada uno con una banda marrón oscura estrecha a lo largo del margen apical, metasoma predominantemente marrón anaranjado o marrón amarillento............... 16
- Tergitos metasómicos con bandas anchas de color marrón oscuronegro, tergitos predominantemente marrón oscuro-negro, especialmente aquellos hacia el ápice del metasoma............ 18

16) Mesonotum uniformemente oscuro naranja-marrón.................
.... *X. (N.) gualanensis*
- Mesonotum anaranjado-marrón oscuro con un triángulo anaranjado invertido en el disco............................... *X. (N.) wilmattae*

17) Tergitos metasómicos primero y segundo marrón amarillento en la base y con una banda marrón oscuro a través de la porción apical, banda marrón oscuro más grande en segundo que el primer tergito; tergitos apicales totalmente marrón oscuro... *X. (N.) nasica*
- Todos los tergitos metasómicos de color marrón amarillento en la base y con banda de color marrón oscuro a través de la porción apical, las bandas de color marrón oscuro se vuelven más anchas en los tergitos apicales, pero los tergitos aún parcialmente marrón amarillento................................... 18

18) Parameres de los genitales masculinos que carecen de un diente ventral en la vista lateral, Fig. 3 f................. *X. (N.) frontalis*
- Parameres de los genitales masculinos con un diente ventral bien definido visible en vista lateral, Fig. 7 g........ *X. (N.) nautlana*

Large Carpenter Bees

Chapter 4: Species Accounts

Xylocopa (Neoxylocopa) columbiensis
Pérez (1901:94, 103)

Figure 1: a: female, dorsal view; **b:** female, lateral view; **c:** female, lateral projection of metatibia; **d:** distribution map in Central America; **e:** front of head, female.

Xylocopa columbiensis Pérez (1901:94, 103), type locality "Chiriquí," three syntype females in collection of the Muséum National d'Histoire Naturelle, Paris.

Female: Overall body length 22-24 mm; metasomal width 9-11 mm; frons with distinct tubercle between antennal bases; integument black; mesosomal disc shining; metasoma densely and finely punctate; metasomal tergites 3-5 with distinct median longitudinal carina; vestiture black; wings hyaline, with feeble green, blue, or violet iridescence; right forewing length 18-23 mm. **Male:** Not examined, but the combination of the bright red dorsal vestiture and the lack of elongate lateral setae on metasomal segments 2-3 should be diagnostic (Pérez 1901:103-104).

Hembra: Longitud total del cuerpo 22-24 mm; ancho de la metasoma 9-11 mm; frons con tubérculo distinto entre las bases antenales; negro del tegumento; disco mesosómico brillante; metasoma denso y finamente punteado; tergitos metasómicos 3-5 con carina longitudinal mediana distinta; pubescencia negra; alas hialinas, con débil iridiscencia de verde, azul, o violeta; longitud de la ala delantera derecha18-23 mm. **Macho:** No examinado, pero la combinación de la vestimenta dorsal roja brillante y la falta de sedas laterales alargadas en los segmentos metasomal 2-3 debería ser diagnóstica (Pérez 1901: 103-104).

Floral Associations: *Hibiscus tiliaceus* L. (Malvaceae); appears to be polylectic, according to Hurd (1978).

Nesting Associations: Unknown (Hurd 1978).

Material Examined: 13 females from the following localities: PANAMÁ: Ancon, Canal Zone; Old Panama; Pan Am Highway, Canal Zone; Panamá City; Taboga Island. Hurd (1978) also records this species from Colombia.

Xylocopa (Neoxylocopa) fimbriata
Fabricius (1804:130)

Figure 2: a: female, dorsal view; **b:** female, front of head, photograph; **c:** female, front of head, line drawing, showing frontal carina and "horns;" **d:** distribution map in México and Central America; **e:** male, dorsal view.

Xylocopa fimbriata Fabricius (1804:340), type locality "America meridionali," holotype female in collection of the Universitetets Zoologiske Museum, Copenhagen.

Xylocopa corniger Westwood (1840:270, pl. 21, f. 3), type locality "unknown," holotype female in collection of the Oxford University Museum, Oxford, England, synonymy by Smith (1874:284).

Xylocopa cornuta Lepeletier (1841:176), type locality "Cayenne," syntype females in collection of the Istituto di Zoologia dell'Universita di Torino, Torino, Italy, synonymy by Erichson (1848:591).

Xylocopa virescens Lepeletier (1841:186), type locality "Cayenne," holotype female in collection of the Istituto di Zoologia dell'Universita di Torino, Torino, Italy, synonymy by Moure (1960:144).

Xylocopa cajennae Lepeletier (1841:203), type locality "Cayenne," syntype male in collection of Oxford University Museum, Oxford, England, synonymy by Smith (1874:284).

Xylocopa fimbriata var. *motaguensis* Cockerell (1912:556), type locality "Gualan, Guatemala," holotype female in collection of American Museum of Natural History, New York.

Female: Overall body length 23-34 mm; metasomal width 13-18 mm; frons with transverse carina and carina with lateral projections or "horns," integument black; mesosoma and metasoma strongly shining; metasomal tergites 4-5 with indistinct median longitudinal carina; vestiture black; wings black with blue or blue-violet iridescence; right forewing length 22-26 mm. **Male:** Overall body length 22-30 mm;

metasomal width 12-14 mm; integument predominantly yellowish-brown, metasoma not banded or otherwise marked with darker pigmentation dorsally; vestiture orange-brown or yellowish-brown; tibiae and ventral surfaces of thorax with dark brownish-black pigmentation; wings transparent, not iridescent; right forewing length 21-24 mm.

Hembra: Longitud total del cuerpo 23-34 mm; anchura de la metasoma 13-18 mm; frons con carina transversal con proyecciones laterales como "cuernos"; negro del tegumento; mesosoma y metasoma fuertemente brillantes; tergitos metasómicos 4-5 con carina longitudinal mediana indistinta; pubescencia negra; alas negras con iridiscencia azul o violeta azulada; longitud de la ala delantera derecha 22-26 mm. **Macho:** Longitud total del cuerpo 22-30 mm; ancho de la metasoma 12-14 mm; tegumento predominantemente marrón amarillento, metasoma sin bandas o marcado con una pigmentación pigmentaria más oscura dorsalmente; pubescencia de color marrón anaranjado o marrón amarillento; tibias y ventrales del tórax con pigmentación negro-pardo oscura; alas transparentes, no iridiscentes; longitud de la ala delantera y derecho 21-24 mm.

Floral Associations: Polylectic floral visitor (Hurd 1978), with records from the following plant species: Bignoniaceae: *Jacaranda* sp., *Tabebuia rosea* DeCandolle. Bixaceae: *Bixa orellana* L., *Cochlospermum vitifolium* (Willdenow) Sprengel. Convolvulaceae: *Turbina corymbosa* (L.) Rafinesque. Cucurbitaceae: *Cucurbita* sp. Fabaceae: *Bauhinia forficata* Link, *Caesalpinia eriostachys* Bentham, *Cajanus cajan* (L.) Millspaugh, *Cassia biflora* L., *Crotalaria incana* L., *Crotalaria* sp., *Dalbergia brownei* Schinz, *Delonix regia* (Bojer ex Hooker) Rafinesque, *Diphysa robinoides* Bentham, *Gliricidia sepium* (Jacquin) Kunth ex Walpers, *Pterocarpus* sp., *Senna bicapsularis* (L.) Roxburgh, *Sesbania herbacea* (Miller) McVaugh. Hamamelidaceae: *Liquidambar styraciflua* L. Fagacaeae: *Quercus* sp. Lamiaceae: *Leonurus sibiricus* L., *Vitex pyramidata* Robinson and Pringle. Lythraceae: *Cuphea ciliata* Ruiz and Pavon, *Heimia salicifolia* Link. Malvaceae: *Hibiscus tiliaceus* L. Marantaceae: *Thalia geniculata* L. Orchidaceae: *Laelia lyonsii* (Lindley) Williams. Polygonaceae: *Antignonon leptopus* Hooker and Arnott. Rubiaceae: *Coffea* sp. Rutaceae: *Zanthoxylum* sp. Sapotaceae: *Manilkara zapota* (L.) Royen. Solanaceae: *Solanum torvum* Swartz (list developed from Hurd 1978 and data labels on specimens examined).

Nesting Associations: Combretaceae: *Laguncularia racemosa* (L.) Gaertner, and undoubtedly other species; often nests in fence posts, rafters, and other structural timbers, causing structural damage (Hurd 1978). Janzen (1966) describes the collapse of a thatched-roof house in a pasture in Oaxaca due to the extensive damage to its structural timbers resulting from the nesting activities of this species.

Material Examined: 184 females and 35 males from the following localities: COSTA RICA: Alajuelo Prov., San Fernando. Guanacaste Prov., Hacienda Taboga; Liberia; Playas del Coco; Santa Rosa National Park. San Jose Prov., 2 km S, 3 km E Desamparados, 1250 m; Escazu; San Ignacio de Acosta; 3 miles S Santa Ana. EL SALVADOR: Quezaltepeque; Santa Tecla. GUATEMALA: 5 miles E Coatepeque; Chicacao; El Rancho; Gualan. MÉXICO: Chiapas: 2 miles N Suchiapa; El Zapotal, 2 miles N Tuxtla Gutierrez; Suchiapa. Colima: Colima; Manzanillo. Distrito Federal: México. Guerrero: 3 miles N Las Cruces. Jalisco: 8 km W Tequila. Morelos: 16 miles S of Cuernavaca; Alpuyeca; Tlaltizapan. Nayarit: 8 miles N Tepic; Acaponeta; Tuxpan, Rio San Pedro. Oaxaca: Candelaria; Temescal. Quintana Roo: Cancun. Sinaloa: 2 miles N Villa Union; 3 km E Mazatlan; 5 miles N Mazatlan; 12 miles S Mazatlan; Cualican. Veracruz: 18 miles SE Alvarado; Cotaxtla; Cotaxtla Experiment Station, Cotaxtla; Tlacotalpan. NICARAGUA: Chinandega; San Marcos. PANAMÁ: Taboga Island. Hurd (1978) also records this species from Belize; Honduras; México: Michoacan, Nuevo Leon, Puebla, and Sonora, as well as numerous countries in South America and the Caribbean.

Taxonomic notes: Cockerell (1912) described *Xylocopa fimbriata* var. *motaguensis* on the basis of a female specimen from Gualan, Guatemala. The only characters given for separation of this form from the nominate form were the bluish-green and blue wing iridescence (stated to be bluer than in specimens that Cockerell studied from Panamá), and the fact that the frontal "horns" were slightly longer than the corresponding structures in Panamanian specimens. Both characters are quite variable in the large species of specimens that I have examined: wing coloration in females is most often blue but can be blue-green or feeble green as noted by Cockerell (1912) in his Panamanian specimens; the length of the frontal "horns" varies greatly and in some specimens they are little more than stout tubercles while in other specimens they project conspicuously out from the frons. In the material that I studied, there do not appear to be distinct geographic patterns in the distribution of either of these characters, and in fact the length of the frontal horns varies considerably within series of specimens which were collected at the same time at the same locality. The specimen described by Cockerell (1912) under the name *X. fimbriata* var. *motaguensis* falls well within what I consider to be the normal range of variation for this species and thus this name must be considered a synonym of *X. fimbriata*.

Xylocopa (Neoxylocopa) frontalis
(Olivier 1789:64)

Figure 3: a: female with orange stripes on metasoma, dorsal view; **b:** female with black metasoma, dorsal view; **c:** female, front of head, line drawing, showing frontal carina and position of ocelli; **d:** male, dorsal view; **e:** distribution map in México and Central America; **f:** lateral view of parameres of male genitalia.

Apis frontalis Olivier (1789:64), type locality "Cayenne," holotype female, current whereabouts unknown and presumed by Hurd (1978) to be lost.

Xylocopa nitens Lepeletier (1841:176), type locality "Cayenne," holotype female, current whereabouts unknown and presumed by Hurd (1978) to be lost, synonymy by Hurd (1978:81).

Xylocopa fasciata Lepeletier (1841:202), type locality "Brésil," holotype male in collection of the Museum National d'Histoire Naturelle, Paris, synonymy by Smith (1874:284).

Xylocopa quadrimaculata Meunier (1892:64), type locality "Quito," current whereabouts unknown (Hurd 1978), synonymy by Hurd (1978:81).

Xylocopa morio callichlora Cockerell (1911:287), type locality "Piura, Peru," holotype female in collection of the American Museum of Natural History, New York, synonymy by Hurd (1978:81).

Xylocopa frontalis var. *coeruleomicans* Enderlein (1913:158), type locality "Brasilien: Espiritu Santo," holotype female in collection of the Academie Polonaise des Sciences, Warsaw, synonymy by Hurd (1978:81).

Xylocopa frontalis var. *viridimicans* Enderlein (1913:158), type locality "Chiriqui," holotype female in collection of the Academie Polonaise des Sciences, Warsaw, synonymy by Hurd (1978:81).

Xylocopa frontalis fabricii Cockerell (1926:658), type locality "French Guiana (Guyane, Maroni)," holotype female in collection of the American Museum of Natural History, New York, synonymy by Hurd (1978:81).

Xylocopa frontalis roseata Cockerell (1926:658), type locality "Ecuador," two syntype females in collection of the American Museum of Natural History, New York, synonymy by Hurd (1978:81).

Xylocopa frontalis trinitatis Cockerell (1926:658), type locality "Trinidad, B. W. I.," three female and unspecified number of male syntypes in collection of American Museum of Natural History, New York, synonymy by Hurd (1978:81).

Xylocopa frontalis var. *purpureipennis* Cockerell (1949:484), type locality "Zamorano," Honduras, holotype female in collection of National Museum of Natural History, Smithsonian Institution, Washington, D. C., synonymy by Hurd (1978:81).

Xylocopa frontalis var. *obscuripennis* Cockerell (1949:484), type locality "Zamorano," Honduras, holotype female in collection of National Museum of Natural History, Smithsonian Institution, Washington, D. C., synonymy by Hurd (1978:82).

Xylocopa americana Prance (1976:238), *nomen nudum.*

Female: Overall body length 24-33 mm; metasomal width 12-19 mm; frons with a distinct median transverse carina interrupted medially by median ocellus; integument black, metasomal tergites sometimes with transverse reddish-orange bands (see Figure 3); mesosoma and metasoma strongly shining; vesture black; wings black with blue, violet, or greenish-gold iridescence; right forewing length 23-27 mm. **Male:** Overall body length 22-33 mm; metasomal width 11-15 mm; integument yellowish-brown; ventral portions of thorax, tibiae, and an apical band on each metasomal tergite dark brownish-black; vestiture orange-brown or yellowish-brown; wings transparent, not iridescent; right forewing length 17-27 mm.

Hembra: Longitud total del cuerpo 24-33 mm; anchura de la metasoma 12-19 mm; frons con una carina transversal mediana distinta interrumpida medialmente por ocelo mediano; negro del tegumentegumento, tergitos metasómicos a veces con bandas transversales rojo-anaranjado (ver Figura 3); mesosoma y metasoma fuertemente brillantes; pubescencia negra; alas negras con iridiscencia azul, violeta o verdosa; Longitud de la ala delantera derecha 23-27 mm. **Macho:** Longitud total del cuerpo 22-33 mm; anchura metasomal 11-15 mm; tegumento amarillento; porciones ventrales de tórax, tibias y una banda apical en cada tergito metasomal de color marrón oscuro-negro; pubescencia de color marrón anaranjado o marrón amarillento; alas transparentes, no iridiscentes; longitud de la ala delantera derecha 17-27 mm.

Floral Associations: Polylectic floral visitor (Hurd 1978), with records from the following flowering plant species: Acanthaceae: *Thunbergia* sp. Bixaceae: *Bixa orellana* L. Fabaceae: *Bauhinia* sp., *Caesalpinia* sp., *Centrosema brasilianum* (L.) Bentham, *Centrosema plumieri* (Turpin ex Persoon) Bentham, *Crotalaria paulina* Schrank, *Dioclea guianensis* Bentham, *Dioclea lasiocarpa* Martius, *Dipteryx odorata* (Aublet) Willdenow, *Inga edulis* Martius, *Gliricidia sepium* (Jacquin) Kunth ex Walpers, *Mucuna urens* (L.) Medikus, *Senna alata* (L.) Roxburgh, *Senna bacillaris* (L.) Irwin and Barneby, *Senna latifolia* (Meyer) Irwin and Barneby, *Senna oblongifolia* (Vogel) Irwin and Barneby. Hamamelidaceae: *Liquidambar styraciflua* L. Lamiaceae: *Vitex odorata* Huber, *Vitex polygama* Chamisso. Lecythidaceae: *Bertholletia excelsa* Humboldt and Bonpland, *Eschweilera decolorans* Sandwith. Malvaceae: *Hibiscus rosa-sinensis* L. Melastomataceae: *Rhynchanthera mexicana* DeCandolle. Myrtaceae: *Orthostemon sellowianus* Berg. Orchidaceae: *Epidendron* sp. cf. *acuminatum* Ruiz and Pavon y Jimenez, *Oncidium onustum* Lindley, *Prosthechea crassilabia* (Poeppig and Endlicher) Carnevali and Ramirez, *Sobralia violacea* Linden ex Lindley. Passifloraceae: *Passiflora edulis* Sims, *Passiflora quadrangularis* L. Rosaceae: *Rosa* sp. Rubiaceae: *Randia aculeata* L. Sapindaceae: *Paullinia pinnata* L. Sapotaceae: *Manilkara zapota* (L.) Royen. Solanaceae: *Solanum lan-*

ciaefolium Jacquin, *Solanum paniculatum* L., *Solanum torvum* Swartz. Verbenaceae: *Stachytarpheta cayennensis* (Richard) Vahl. Vochysiaceae: *Salvertia convallariaeodora* Saint-Hilaire. Zingiberaceae: *Zingiber* sp. (Hurd 1978 and labels on specimens examined).

Nesting Associations: Apocynaceae: *Haemadictyon* sp. Arecaceae: *Cocos nucifera* L. Lauraceae: *Phoebe porosa* (Nees and Martius) Mez. Malvaceae: *Theobroma cacao* L. Myrtaceae: *Eucalyptus* sp. (Hurd 1978). Unlike *X. (N.) fimbriata*, this species does not appear to excavate nests in structural timbers (Hurd (1978).

Material Examined: 60 females and 21 males from the following localities: COSTA RICA: Cartago Province: 2.5 km E and 4 km N Chitaria. Guanacaste Province: 2 miles W Arenal. Heredia Province: Finca de la Selva. San José Province: 1 mile ESE San Isidro, 703 m; 4.5 miles ESE San Isidro de General; San Ignacio de Acosta; San Ysidro. EL SALVADOR: 2.5 miles W Quezaltepeque; Los Chorros; Mt. San Salvador; Santa Tecla. GUATEMALA: 5 miles E Coatepeque; Coatepeque, Quezaltenango; Guatemala City, 5000 feet; Lake Atitlán, 6000 feet; Variedades, 500 feet. HONDURAS: 20 miles from Tegucigalpa; Zamorano. MÉXICO: Chiapas: 20 miles NW Francia; 25 miles S Tuxla Gutierrez; 15 miles N Comitán; El Zapotal, 2 miles S. Tuxla Gutierrez; Ocozocoautla; Simojovel; Suchiapa. Distrito Federal: no locality specified. Veracruz: Córdoba. PANAMÁ: Albrook Field; Ancón, Canal Zone; Barro Colorado Island, Canal Zone; Cerro Campana, steep hillside forest, 3.6 km above Campana; Campana-Chica Rd., Panamá Prov. Cerro Cedro; Chama, Panamá Prov.; Chiriquí (1 female); Culebra-Arrijon Trail, Canal Zone; El Volcan, Chiriquí; Gatún, Canal Zone. Hurd (1978) also records this species from Belize and Nicaragua, as well as numerous countries in South America and the Caribbean.

Xylocopa (Neoxylocopa) gualanensis
Cockerell (1912:555)

Figure 4: a: female, dorsal view; **b:** male, dorsal view; **c:** female, lateral view; **d:** female, lateral projection of metatibia; **e:** distribution map; **f:** female, front of head.

Xylocopa wilmattae gualanensis Cockerell (1912:555), type locality "Gualan, Guatemala," holotype female in collection of the American Museum of Natural History, New York.

Xylocopa gualanensis Cockerell (Hurd and Moure 1963:151; Hurd 1978:62).

Female: Overall body length 21-27 mm; metasomal width 10-13 mm; frons with a small but distinct tubercle between the antennal bases; integument black; mesosoma shining, sparsely punctate; metasoma densely and finely punctate; metasomal tergites 4-5 with distinct median longitudinal carina; vestiture black; wings pigmented black, with bluish-green iridescence with violet tints throughout, not strongly violet; right forewing length 18-20 mm. **Male:** Overall body length 22-24 mm; metasomal width 9-11 mm; integument predominantly orange-brown or yellowish-brown, each metasomal tergite with a narrow apical band of dark brownish-black pigmentation; vestiture yellowish-brown; wings transparent, not iridescent; right forewing length 17-19 mm.

Hembra: Longitud total del cuerpo 21-27 mm; ancho de la metasoma 10-13 mm; frons con un pequeño pero distinto tubérculo entre las antenas; negro del tegumento; mesosoma brillante, poco punteado; metasoma denso y finamente punteado; tergitos metasómicos 4-5 con carina longitudinal mediana distinta; pubescencia negra; alas pigmentadas de color negro, con iridiscencia azulado-verde con tonos violeta en todo, no muy violeta; longitud de la ala delantera derecha 18-20 mm. **Macho:** Longitud total del cuerpo 22-24 mm; ancho de la metasoma 9-11 mm; tegumento predominantemente marrón anaranjado o marrón amarillento, cada tergito metasomal con una banda apical estrecha de pigmentación pardo-negro oscuro; pubescencia marrón amarillento; alas transparentes, sin iridiscencia; longitud de la ala delantera derecha 17-19 mm.

Floral Associations: Polylectic floral visitor (Hurd 1978), with records from the following plant species: Fabaceae: *Calliandra portoricensis* (Jacquin) Bentham, *Canavalia rosea* (Swartz) DeCandolle, *Cassia biflora* L., *Crotalaria retusa* L., *Delonix regia* (Bojer ex Hooker) Rafinesque. Plantaginaceae: *Stemodia durantifolia* (L.) Swartz. Polygonaceae: *Antigonon leptopus* Hooker and Arnott. Solanaceae: *Solanum* nr. *donnell-smithii* Coult (list developed from Hurd 1978 and from labels on specimens examined).

Nesting Associations: dead branches of *Calliandra portoricensis* (Jacquin) Bentham (Fabaceae) and also in structural timbers (Hurd 1978).

Material Examined: 80 females and 15 males from the following localities: COSTA RICA: Guanacaste: 6 miles S, 6 miles E Canas, Taboga; 14 km SW Liberia, via Highway 21; Finca Pacifica; Playas del Coco; Santa Rosa National Park; Puntarenas: Osa Peninsula. EL SALVADOR: Santa Tecla. GUATEMALA: El Rancho. HONDURAS: Greater Swan Island. NICARAGUA: Madriz Prov., Somoto.

Notes: This species was originally described by Cockerell (1912) as a subspecies of *X. (N.) wilmattae* based on differences in the wing iridescence and the shape of the metatibial process. It was treated as a full species by Hurd and Moure (1963) and Hurd (1978). Females and males of this species are very similar to those of *X. (N.) wilmattae*, and future research may show that these two taxa are actually synonyms. As discussed below under *X. (N.) wilmattae*, the name *X. (N.) wilmattae* would arguably have priority over *X. (N.) gualanensis*, even though both names were published on the same date on the same page of the same article in the same journal.

Xylocopa (Neoxylocopa) mexicanorum
Cockerell (1912:555)

Figure 5: a: female, dorsal view; **b:** male, dorsal view; **c:** female, lateral view; **d:** female, lateral projection of metatibia; **e:** distribution map in México; **f:** female, front of head.

Xylocopa mexicanorum Cockerell (1912:555), type locality "Rio Nautla, Vera Cruz, México," two females syntypes in collection of the American Museum of Natural History, New York.

Female: Overall body length 20-26 mm; metasomal width 9-13 mm; frons with a small but distinct tubercle between the antennal bases; integument black; mesosoma shining, with small, scattered punctures; metasoma matte, with small, scattered punctures; metasomal tergites 3-5 with or without median longitudinal carina; vestiture black; wings pigmented black, with strong violet iridescence and blue or bluish-green tints; right forewing length 19-22 mm. **Male:** Overall body length 24-28 mm; metasomal width 10-13 mm; integument predominantly orange-brown or yellowish-brown; ventral portion of thorax and tibiae darker brown; vestiture orange-brown; wings transparent, not iridescent; right forewing length 18-20 mm.

Hembra: Longitud total del cuerpo 20-26 mm; anchura de la metasoma 9-13 mm; frons con un pequeño pero distinto tubérculo entre las antenas; negro del tegumento; mesosoma brillante, con pequeñas punciones dispersas; metásoma mate, con pequeñas punciones dispersas; tergitos metasómicos 3-5 con o sin carina longitudinal media; pubescencia negra; alas negro pigmentado, con fuerte iridiscencia violeta y tonos azules o verde azulado; longitud de la ala delantera derecha 19-22 mm. **Macho:** Longitud total del cuerpo 24-28 mm; ancho de la metasoma 10 a 13 mm; tegumento predominantemente marrón anaranjado o marrón amarillento; porción ventral del tórax y de las tibias marrón oscuro; pubescencia naranja-marrón; alas transparentes, no iridiscentes; longitud de la ala delantera derecha18-20 mm.

Floral Associations: Polylectic floral visitor (Hurd 1978), with records from the following plant species: Bignoniaceae: *Tecoma stans* (L.) Jussieu ex Kunth. Fabaceae: *Prosopis* sp. Zygophyllaceae: *Larrea tridentata* (DeCandolle) Coville (Hurd 1978).

Nesting Associations: Unknown (Hurd 1978).

Material Examined: 12 females and 5 males from the following localities: HONDURAS: Swan Island. MÉXICO: Jalisco: Chamela; Nuevo Leon: Linares; Michoacan: Contepec, near Jalapa; Oaxaca: Tehuantepec; San Luis Potosí: Valles; Tamaulipas: El Limon. I also examined specimens from Texas in the United States. Hurd (1978) also records this species from México: Guerrero, Veracruz; USA: eastern Arizona, New Mexico.

Xylocopa (Neoxylocopa) nasica
Pérez (1901:91, 102)

Figure 6: a: female, dorsal view; **b:** female, lateral projection of metatibia, line drawing; **c:** female, front view of head; **d:** male, dorsal view; **e:** distribution map in Central America.

Xylocopa nasica Pérez (1901:91, 102), type locality "Chiriquí," holotype female in collection of the Muséum National d'Histoire Naturelle, Paris.

Female: Overall body length 26-28 mm; metasomal width 13-14 mm; clypeus and frons tumescent and elevated medially; integument black; mesosoma shining, with small indistinct punctures; metasoma matte, with dense small scattered punctures; metasomal tergites 2-5 with more or less distinct median carina; vestiture black; wings dark hyaline brown, with strong greenish-gold iridescence; right forewing length 23-24 mm. **Male:** Overall body length 23-28 mm; metasomal width 11-13 mm; integument yellowish-brown and orange brown; ventral portions of thorax and tibiae darker brown; metasomal tergites 1-2 yellowish-brown at base with a darker brown band along apical margin; remaining metasomal tergites dark brown; vestiture orange-brown with darker brown setae on metasomal disc and apex; wings transparent, not iridescent; right forewing length 20-21 mm.

Hembra: Longitud total del cuerpo 26-28 mm; anchura de la metasoma 13-14 mm; clypeus y frons tumescentes y elevados medial; negro del tegumento; mesosoma brillante, con pequeñas punciones indistintas; metásoma mate, con pequeñas y densas punciones dispersas; tergitos metasómicos 2-5 con carina mediana más o menos clara; pubescencia negra; alas de color marrón hialino oscuro, con fuerte iridiscencia de color verdoso; longitud de la ala delantera derecha 23-24 mm. **Macho:** Longitud total del cuerpo 23-28 mm; anchura de la metasoma 11-13 mm; integumento marrón amarillento y marrón anaranjado; porciones ventrales de tórax y tibias de color marrón oscuro; tergitos metasómicos 1-2 marrón amarillento en la base con una banda marrón oscura a lo largo del margen apical; tergitos metasómicos restantes marrón oscuro; pubescencia de color naranja-marrón con setas marrones más oscuras en el disco metasomal y ápice; alas transparentes, no iridiscentes; longitud de la ala delantera derecha 20-21 mm.

Floral Associations: Unknown (Hurd 1978).

Nesting Associations: Unknown (Hurd 1978).

Materials Examined: 5 females and 4 males from the following localities: COSTA RICA: Corcovado, Osa. PANAMÁ: Alhajuelo, Canal Zone; Frijoles, Canal Zone; La Chorrera; Porto Bello Trail, Continental Divide; Red Tank; Taboga, Canal Zone. I also examined a female specimen from Colombia.

Xylocopa (Neoxylocopa) nautlana
Cockerell (1904:29)

Figure 7: a: female, dorsal view; **b:** female, front view of head, photograph; **c:** female, front view of head, line drawing, showing frontal carina and position of ocelli; **d:** female, lateral view; **e:** distribution map; **f:** male, dorsal view; **g:** lateral view of parameres of male genitalia, with ventral projection indicated by arrow.

Xylocopa nautlana Cockerell (1904:29), type locality "Rio Nautla, in the neighborhood of San Rafael, State of Vera Cruz, México," holotype female in collection of the California Academy of Sciences, San Francisco.

Female: Overall body length 23-30 mm; metasomal width 9-14 mm; frons with central transverse carina, carina entire, with median ocellus located above carina; integument black, mesosoma and metasoma strongly shining; vestiture black; wings black, with brilliant blue-violet iridescence; right forewing length 21-25 mm. **Male:** Overall body length 24-29 mm; metasomal width 13-20 mm; integument yellowish-brown and orange brown; ventral portions of thorax and tibiae dark brown; each metasomal tergite yellowish-brown at base but with a darker brown band across apex; vestiture yellowish-brown, with darker brown setae at apex of abdomen; wings transparent, not iridescent; right forewing length 24-28 mm.

Hembra: Longitud total del cuerpo 23-30 mm; anchura de la metasoma 9-14 mm; frons con carina transversa central, carina entera, con ocelo mediano situado sobre carina; negro del tegumento, mesosoma y metasoma fuertemente brillante; pubescencia negra; alas negras, con brillante iridiscencia azul-violeta; longitud de la ala delantera y derecho 21-25 mm. **Macho**: Longitud total del cuerpo 24-29 mm; anchura de la metasoma 13-20 mm; integumento marrón amarillento y marrón anaranjado; porciones ventrales de tórax y tibias de color marrón oscuro; cada tergito metasomal marrón amarillento en la base pero con una banda marrón oscura a través del ápice; pubescencia marrón amarillento, con setas marrones más oscuras en el ápice del abdomen; alas transparentes, no iridiscentes; longitud de la ala delantera y derecho 24-28 mm.

Floral Associations: Oleaceae: *Jasminum* sp. Probably a polylectic floral visitor, but the floral associations of this species are poorly known (Hurd 1978).

Nesting Associations: Unknown (Hurd 1978)

Materials Examined: 10 females and 5 males from the following localities: COSTA RICA: Guanacaste Province: 2 miles W Arenal. MÉXICO: Chiapas: Simojovel. Oaxaca: Valle Nacional. Quintana Roo: Filipe Carrilo Puerto. San Luis Potosí: Tamazunchale. Tabasco: Frontera. Tamaulipas: Tampico. Veracruz: Coatepec. Hurd (1978) also records this species from Guatemala; Honduras; Mexico: Campeche, Hidalgo, Yucatan. Based on the known distribution map (Fig. 7 e) the species likely also occurs in Nicaragua and possibly Belize.

Notes: Males and females of this species are extremely similar to those of *X. (N.) frontalis*. However, clear and consistent differences in the structure of the frontal carina in the females and the structure of the male genitalia (see keys and illustrations) indicate that these are actually separate species. Females of these species can be separated by the structure of the frontal carina and the placement of the ocelli, as indicated in the key, illustrations, and diagnoses, while males of the two species have subtle differences in the shape of the reproductive structures, particularly the parameres (see illustrations). Any material of *X. (N.) frontalis* from Central America should be examined for individuals of this species, as this species is almost certainly overlooked in many collections where it is often misidentified as *X. (N.) frontalis*. Careful examination of additional material could significantly expand the known distribution of this species.

Xylocopa (Neoxylocopa) ocellaris
Pérez (1901:90)

Figure 8: a: female, dorsal view; **b:** female, lateral view; **c:** female, front of head; **d:** female, lateral projection of metatibia; **e:** distribution map in Central America.

Xylocopa ocellaris Pérez (1901:90), type locality "Chiriquí," holotype female in collection of the Muséum National d'Histoire Naturelle, Paris.

Female: Overall body length 22-25 mm; metasomal width 11-12 mm; frons broader and lacking the distinctive tubercle between antennal bases; integument black; mesosoma shining; metasoma matte, finely and distinctly punctate; metasomal tergites 3-5 with more or less distinct median carina; vesture reddish-brown on vertex of head, anterior and posterior portions of mesosoma, and apex of abdomen; otherwise black; wings dark hyaline with feeble golden-green iridescence basally and feeble golden-red iridescence apically; right forewing length 21-23 mm. **Male:** Unknown.

Hembra: Longitud total del cuerpo 22-25 mm; anchura de la metasoma 11-12 mm; frons más amplios y sin el distintivo tubérculo entre las bases antenales; negro del tegumento; mesosoma brillante; metásoma mate, finamente y distintamente punteado; tergitos metasómicos 3-5 con carina mediana más o menos clara; pubescencia de color marrón rojizo en el vértice de la cabeza, porciones anterior y posterior del mesosoma y ápice del abdomen, de lo contrario negro; alas húmedas oscuras con débil iridiscencia de color dorado-verde basalmente y débil iridiscencia rojo dorado apicalmente; longitud de la ala delantera derecha 21-23 mm. **Macho:** Desconocido.

Floral Associations: Unknown (Hurd 1978).
Nesting Associations: Unknown (Hurd 1978).
Materials Examined: 5 females from the following locality: PANAMÁ: Taboga Island. Hurd (1978) also records this species from Colombia.
Notes: The reddish-brown to dark reddish-brown vesture on the frons and mesosoma is an important diagnostic character for females of this species; this vesture is generally black in related species. The reddish-brown coloration of the vesture is very striking in the female

Large Carpenter Bees

holotype of this species, which has recently been illustrated using high-resolution digital photographs on the website of the Muséum National d'Histoire Naturelle, Paris. Other females such as the specimen illustrated have darker vestiture which is however still noticeably reddish-brown. The pigmentation and iridescence of the wings is also helpful in diagnosing females of this species, particularly those individuals which have darker vestiture on the head and mesosoma.

Xylocopa (Neoxylocopa) varipuncta
Patton (1879:60)

Figure 9: a: female, dorsal view; **b:** male, dorsal view; **c:** female, lateral view; **d:** female, lateral projection of metatibia; **e:** distribution map in México; **f:** female, front view of head.

Xylocopa varipuncta Patton (1879:60), type locality "Arizona," two female syntypes, current whereabouts unknown and (according to Hurd 1978) apparently lost.

Female: Overall body length 19-24 mm; metasomal width 9-10 mm; frons with a small tubercle between antennal bases; integument black; vestiture black; mesosomal disc shining; metasomal tergites matte; wings hyaline at base, apical field darker, with brilliant violet-gold iridescence at base, apical field with brilliant reddish-gold iridescence; right forewing length 18-20 mm. **Male:** Overall body length 20-27 mm; metasomal width 8-11 mm; integument yellowish-brown or orange-brown; vestiture yellowish-brown to orange; metasomal tergites and most of mesosoma except central bare patch with dense vestiture; central bare patch of mesosoma shining; wings hyaline, not iridescent; veins brownish-black; right forewing length 16-19 mm.

Hembra: Longitud total del cuerpo 19-24 mm; anchura de la metasoma 9-10 mm; frons con un pequeño tubérculo entre las bases antenales; negro del tegumento; pubescencia negra; disco mesosómico brillante; tergitos metasómicos mate; alas hialinas en la base, campo apical más oscuro, con brillante iridiscencia violeta-dorada en la base, campo apical con brillantes iridescencias rojizo-doradas; longitud de la ala delantera derecha 18-20 mm. **Macho:** Longitud total del cuerpo 20-27 mm; anchura de la metasoma 8-11 mm; integumento marrón amarillento o marrón anaranjado; pubescencia de color amarillento a naranja; tergitos metasómicos y la mayor parte del mesosoma excepto el parche desnudo central con vestimenta densa; parche desnudo central de mesosoma brillante; alas hialinas, no iridiscentes; venas marrón-negro; longitud de la delantera derecha 16-19 mm.

Floral Associations: Polylectic floral visitor (Hurd 1978), with records from the following plant species: Apocynaceae: *Asclepias* sp. Asteraceae: *Coreopsis* sp. Brassicaceae: *Brassica rapa* L. Caprifoliaceae: *Lonicera japonica* Thunberg. Cleomaceae: *Peritoma arborea* (Nuttall) Iltis, *Wislizenia mammilata* Rose, *Wislizenia refracta* Engelmann. Cucurbitaceae: *Cucurbita argyrosperma* Koch, *Cucurbita foetidissima* Kunth in von Humboldt, *Cucurbita maxima* Duchesne, *Cucurbita moschata* Duchesne ex Poiret, *Cucurbita pepo* L. Fabaceae: *Acacia* sp., *Astragalus douglasii* var. *parishii* (Gray) Jones, *Eysenhardtia polystachya* (Ortega) Sargent, *Lathyrus odoratus* L., *Lupinus paynei* Davidson, *Medicago sativa* L., *Parkinsonia aculeata* L., *Parkinsonia florida* (Bentham ex Gray) Watson, *Prosopis glandulosa* var. *torreyanum* (Benson) Johnston, *Sesbania herbacea* (Miller) McVaugh, *Vigna unguiculata* (L.) Walpers, *Wisteria* sp. Lamiaceae: *Salvia* sp., *Trichostema lanceolatum* Bentham. Malvaceae: *Gossypium herbaceum* L., *Sphaeralcea emoryi* Torrey ex Gray. Martyniaceae: *Proboscidea althaeifolia* (Bentham) Decaisne. Onagraceae: *Oenothera elata* Kunth. Papaveraceae: *Argemone* sp., *Eschscholzia californica* Chamisso. Passifloraceae: *Passiflora* sp. Plantaginaceae: *Keckiella antirrhinoides* (Bentham) Straw. Rhamnaceae: *Ceanothus velutinus* var. *hookeri* Johnston. Rosaceae: *Cotoneaster* sp. Scrophulariaceae: *Buddleia* sp. Solanaceae: *Datura inoxia* Miller, *Solanum douglasii* Dunal, *Solanum elaeagnifolium* Duval, *Solanum lysopersicum* L. Verbenaceae: *Lantana camara* Miller. Zygophyllaceae: *Larrea tridentata* (DeCandolle) Colville (Hurd 1978 and labels on specimens examined).

Nesting Associations: Anacardiaceae: *Schinus molle* L. Apocynaceae: *Nerium oleander* L. Asparagaceae: *Yucca* sp. Betulaceae: *Alnus rhombifolia* Nuttal. Cactaceae: *Stenocereus thurberi* (Engelmann) Buxbaum. Fagaceae: *Quercus agrifolia* Née. Juglandaceae: *Juglans* regia L. Malvaceae: *Ochroma lagopus* Swartz. Myrtaceae:

Eucalyptus sp. Rosaceae: *Malus* sp., *Prunus armeniaca* L. Salicaceae: *Populus* sp. Thymelaeaceae: *Edgeworthia chrysantha* Siebold and Zuccarini. Most nests are excavated in decaying or rotting wood, occasionally in structural timber (Hurd 1978).

Materials Examined: 24 females and 2 males from the following localities: MÉXICO: Baja California: 2 miles S La Paz; 2 miles NE Cabo San Lucas; Isla Cerralvo, Ruffo Ranch; San Antonio; Santo Domingo. Michoacan: Zamora. Sinaloa: 35 miles N de Los Mochis. Sonora: 60 km NE of Alamos; Ciudad Obregon; 16 miles NE Ciudad Obregon; 15 miles E Navojos; Nogales. I also examined specimens from Arizona and California in the United States. Hurd (1978) also records this species from México: Colima, Durango, Guanajuato, Guerrero, Hidalgo, Jalisco, Morelos, Nayarit, Puebla, San Luis Potosí, Zacatecas; USA: southern Nevada, western New Mexico.

Notes: Considerable debate, reviewed by Hurd (1978) has focused on whether this species is conspecific with the Hawaiian form described by Smith (1874) under the name *X. sonorina* Smith (1874:278). Females and males of these two forms are quite similar, although the same can also be said for many other species of subgenus *Neoxylocopa*. Hurd (1978) kept the two forms separate based on the available morphological evidence, including dissections of male reproductive structures which were illustrated by Hurd and Moure (1963). Molecular systematics approaches such as DNA barcoding or possibly some of the newer genomics techniques could be applied to help resolve this issue, as well as the related question of whether the Hawaiian form represents a recent anthropogenic introduction or whether the ancestors of the current Hawaiian populations arrived prior to European settlement. If the two forms are ever shown to be conspecific, the name *X. sonorina* would have priority.

Xylocopa (Neoxylocopa) wilmattae
Cockerell (1912:555)

Figure 10: a: female, dorsal view; **b:** male, dorsal view; **c:** female, lateral view; **d:** female, lateral projection of metatibia; **e:** distribution map; **f:** front view of head.

Xylocopa wilmattae Cockerell (1912:555), type locality "Guatemala City, Guatemala," holotype female in collection of the American Museum of Natural History, New York.

Female: Overall body length 23-28 mm; metasomal width 7-8 mm; integument black, mesosoma and metasoma matte, feebly shining; vestiture black; wings dark brownish-black, with feeble blue-green and blue-violet iridescence; right forewing length 20-22 mm; metatibial lateral process with apices bifid, each lobe roughly equal in size. **Male:** Overall body length 26 mm; metasomal width 12 mm; integument orange-brown, legs and venter of metasoma yellowish-brown, mesonotum with a median triangular orange marking as described in key; wings transparent, not at all iridescent; right forewing length 22 mm.

Hembra: Longitud total del cuerpo 23-28 mm; Ancho de la metasoma 7-8 mm; Tegumento negro, mesosoma y metasoma mate, débilmente brillante; Puibescencia negro; Alas de color marrón oscuro-negro, con una débil irididscencia azul-verde y azul-violeta; Longitud de la ala delantera derecha 20-22 mm; Proceso metatibial lateral con ápices bífidos, cada lóbulo aproximadamente igual en tamaño. **Macho:** Longitud total del cuerpo 26 mm; Ancho de los metastásicos 12 mm; Tegumento naranja-marrón, piernas y ventre de metasoma marrón amarillento, mesonotum con una marca anaranjada triangular mediana como se describe en clave; Alas transparentes, sin iridiscencia; Longitud de la ala delantera derecha 22 mm.

Floral Associations: Unknown (Hurd 1978).

Nesting Associations: Unknown (Hurd 1978).

Materials Examined: 5 females and 1 male from the following localities: GUATEMALA: Guatemala City, Selola Prov., 10 miles E Panajachel. HONDURAS: Zamorano.

Notes: Specimens identified by Cockerell as this species are very similar to those of *X. (N.) gualanensis* and may in fact be conspecific. Both of these names were published by the same author on the same date and on the same page (Cockerell 1912:555). However, the name *X. wilmattae* appears on the page above the name *X. gualanensis*, and the name *X. gualanensis* was originally published as a subspecies of *X. wilmattae*. Thus, the name *X. wilmattae* would arguably be considered to have priority over *X. gualanensis*.

Figure Captions

Figure 1: *Xylocopa columbiensis:* **a:** female, dorsal view; **b:** female, lateral view; **c:** female, lateral projection of metatibia; **d:** distribution map in Central America; **e:** front of head, female.

Figure 2: *Xylocopa fimbriata:* **a:** female, dorsal view; **b:** female, front of head, photograph; **c:** female, front of head, line drawing, showing frontal carina and "horns;" **d:** distribution map in México and Central America; **e:** male, dorsal view.

Figure 3: *Xylocopa frontalis:* **a:** female with orange stripes on metasoma, dorsal view; **b:** female with black metasoma, dorsal view; **c:** female, front of head, line drawing, showing frontal carina and position of ocelli; **d:** male, dorsal view; **e:** distribution map in México and Central America; **f:** lateral view of parameres of male genitalia.

Figure 4: *Xylocopa gualanensis:* **a:** female, dorsal view; **b:** male, dorsal view; **c:** female, lateral view; **d:** female, lateral projection of metatibia; **e:** distribution map; **f:** female, front of head.

Figure 5: *Xylocopa mexicanorum:* **a:** female, dorsal view; **b:** male, dorsal view; **c:** female, lateral view; **d:** female, lateral projection of metatibia; **e:** distribution map in México; **f:** female, front of head.

Figure 6: *Xylocopa nasica:* **a:** female, dorsal view; **b:** female, lateral projection of metatibia; **c:** female, front view of head; **d:** male, dorsal view; **e:** distribution map in Central America.

Figure 7: *Xylocopa nautlana:* **a:** female, dorsal view; **b:** female, front view of head, photograph; **c:** female, front view of head, line drawing, showing frontal carina and position of ocelli; **d:** female, lateral view; **e:** distribution map; **f:** male, dorsal view; **g:** male genitalia, lateral view, with ventral projection indicated by arrow.

Figure 8: *Xylocopa ocellaris:* **a:** female, dorsal view; **b:** female, lateral view; **c:** female, front of head; **d:** female, lateral projection of metatibia; **e:** distribution map in Central America.

Figure 9: *Xylocopa varipuncta:* **a:** female, dorsal view; **b:** male, dorsal view; **c:** female, lateral view; **d:** female, lateral projection of metatibia; **e:** distribution map in México; **f:** female, front view of head.

Figure 10: *Xylocopa wilmattae:* **a:** female, dorsal view; **b:** male, dorsal view; **c:** female, lateral view; **d:** female, lateral projection of metatibia; **e:** distribution map; **f:** front view of head.

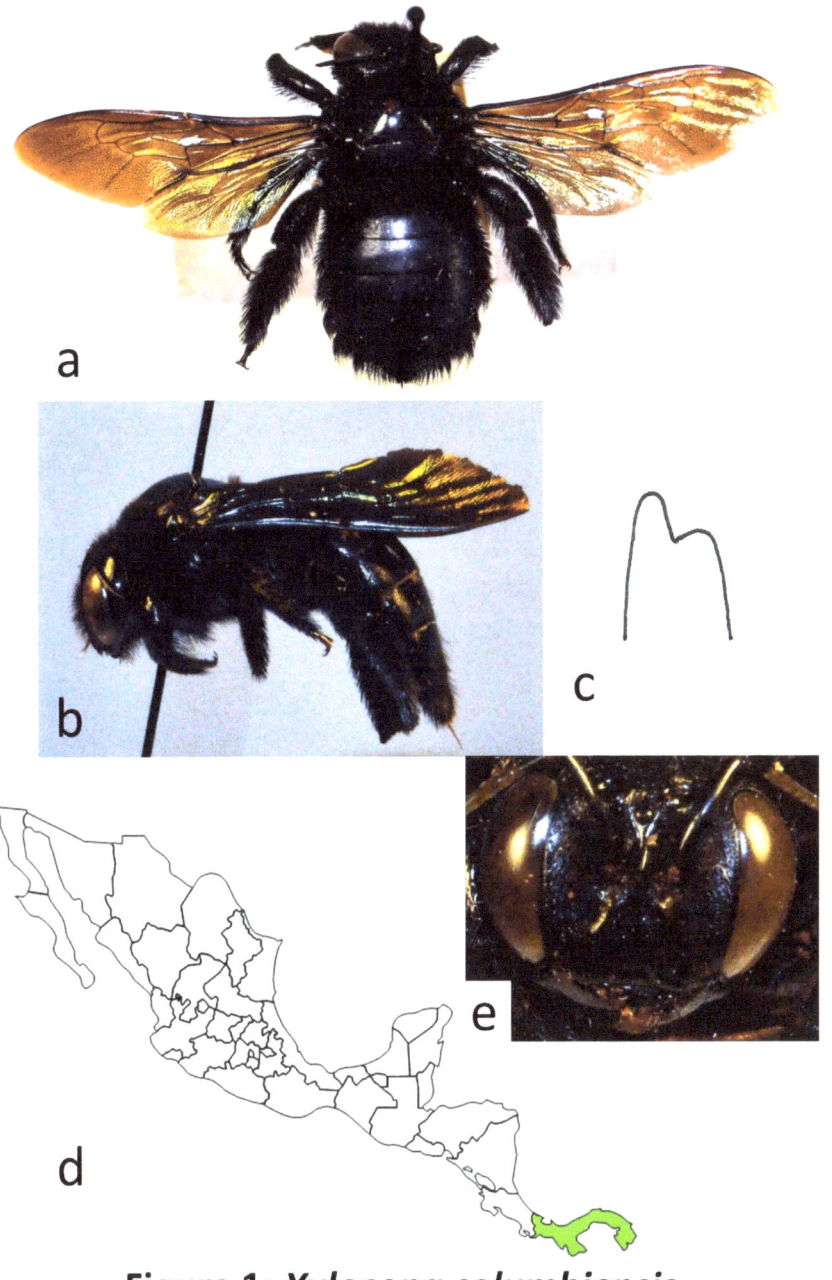

Figure 1: *Xylocopa columbiensis*

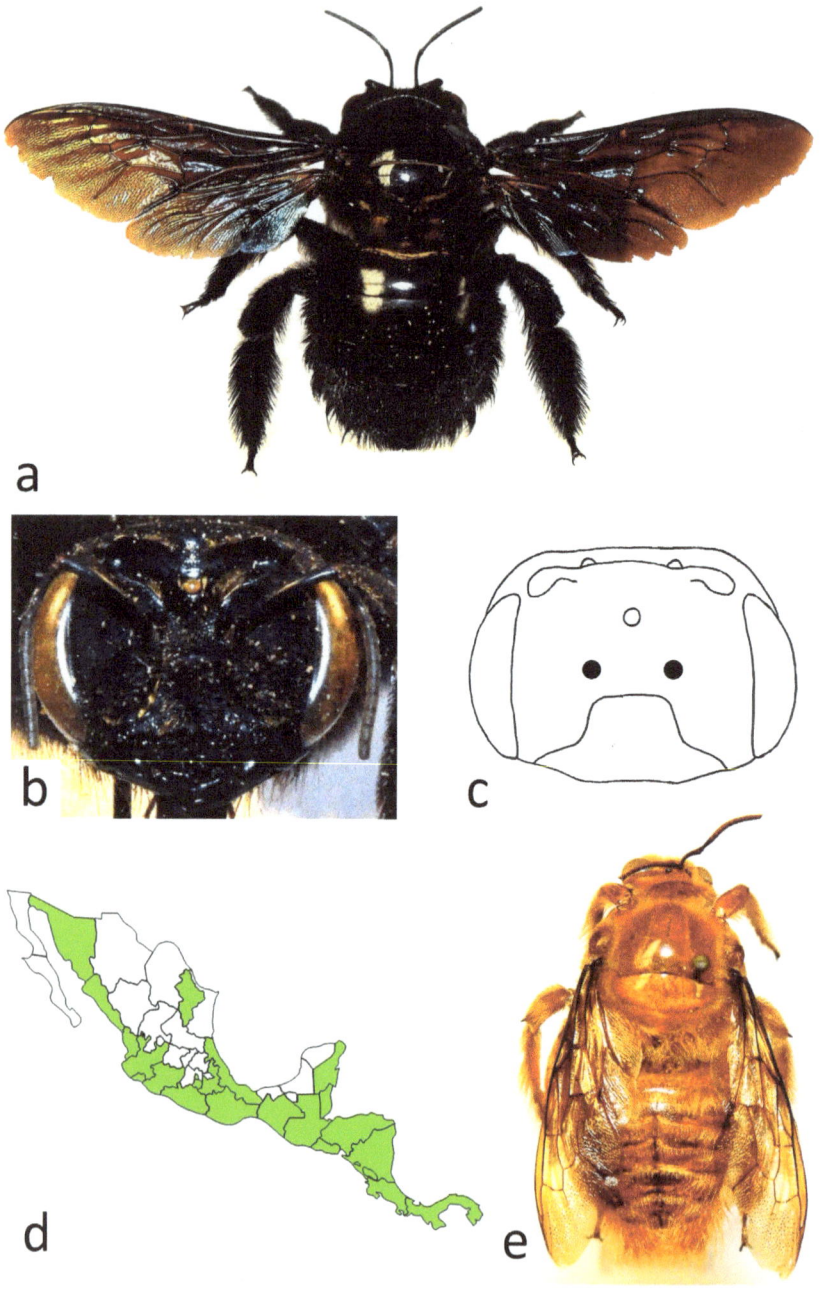

Figure 2: *Xylocopa fimbriata*

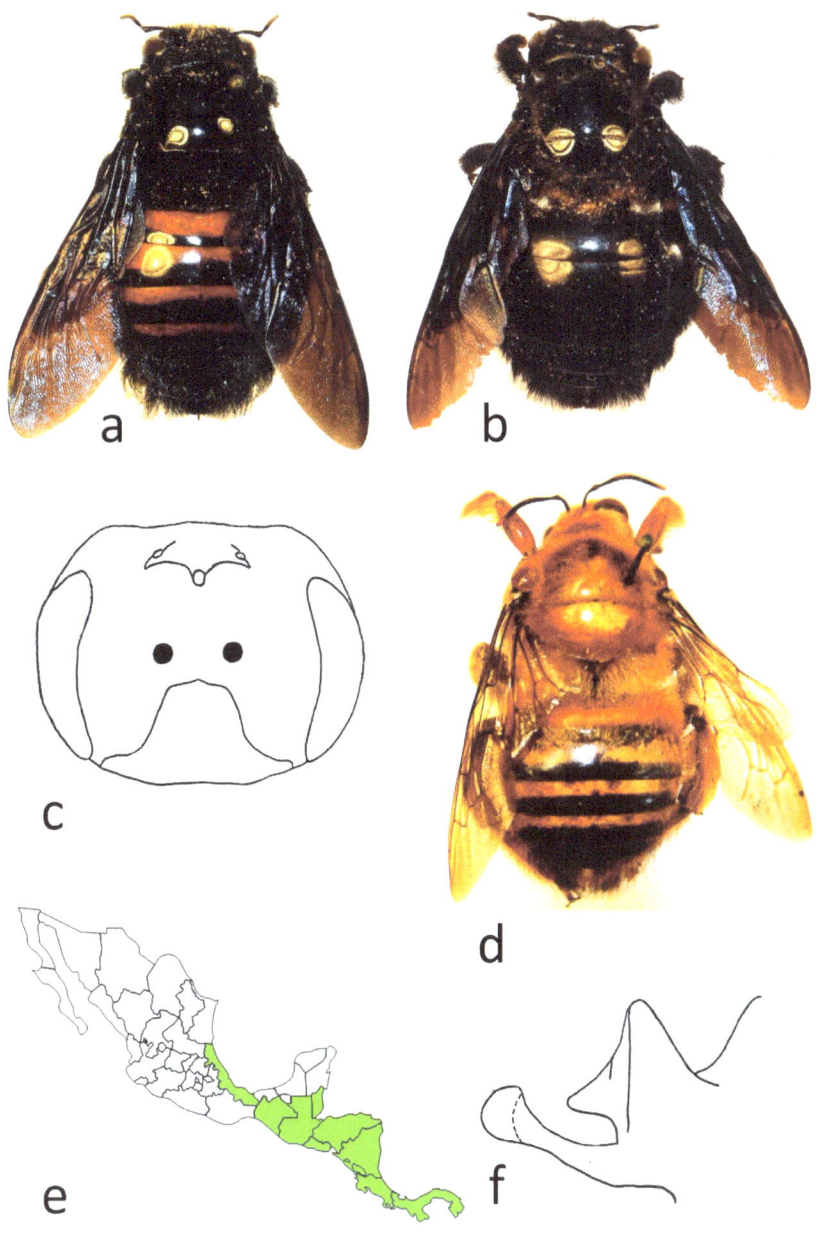

Figure 3: *Xylocopa frontalis*

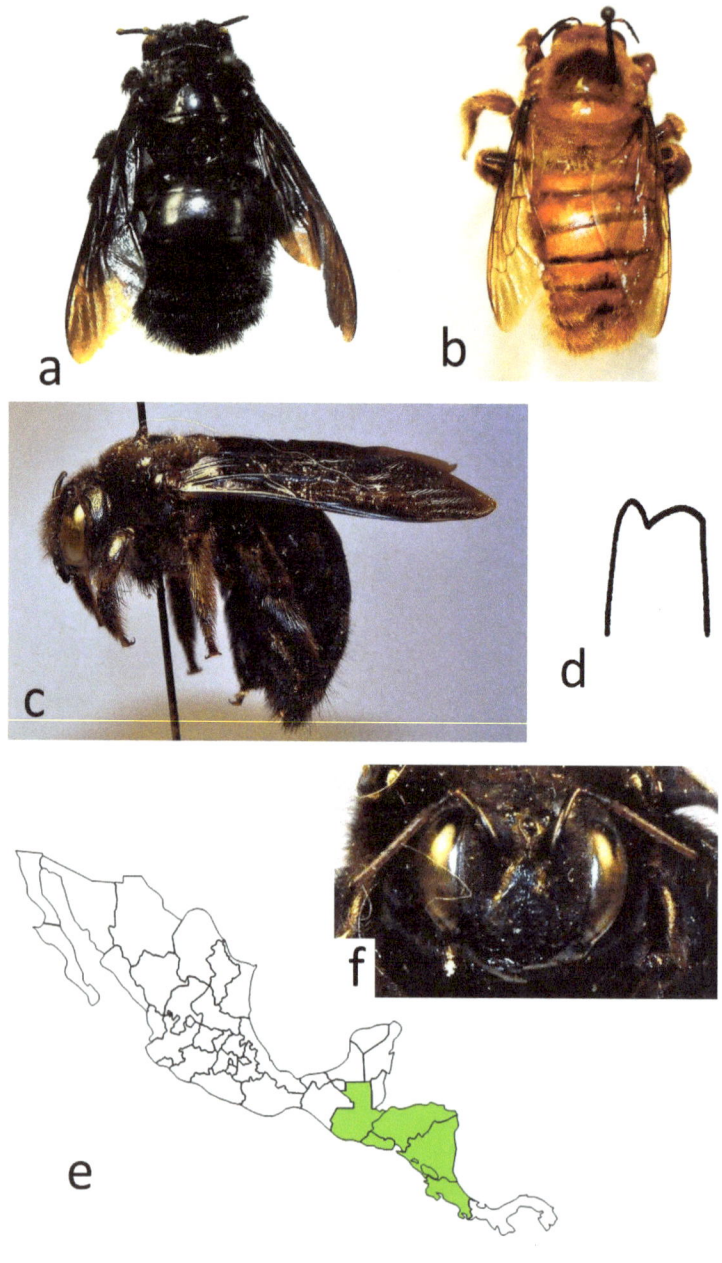

Figure 4: *Xylocopa gualanensis*

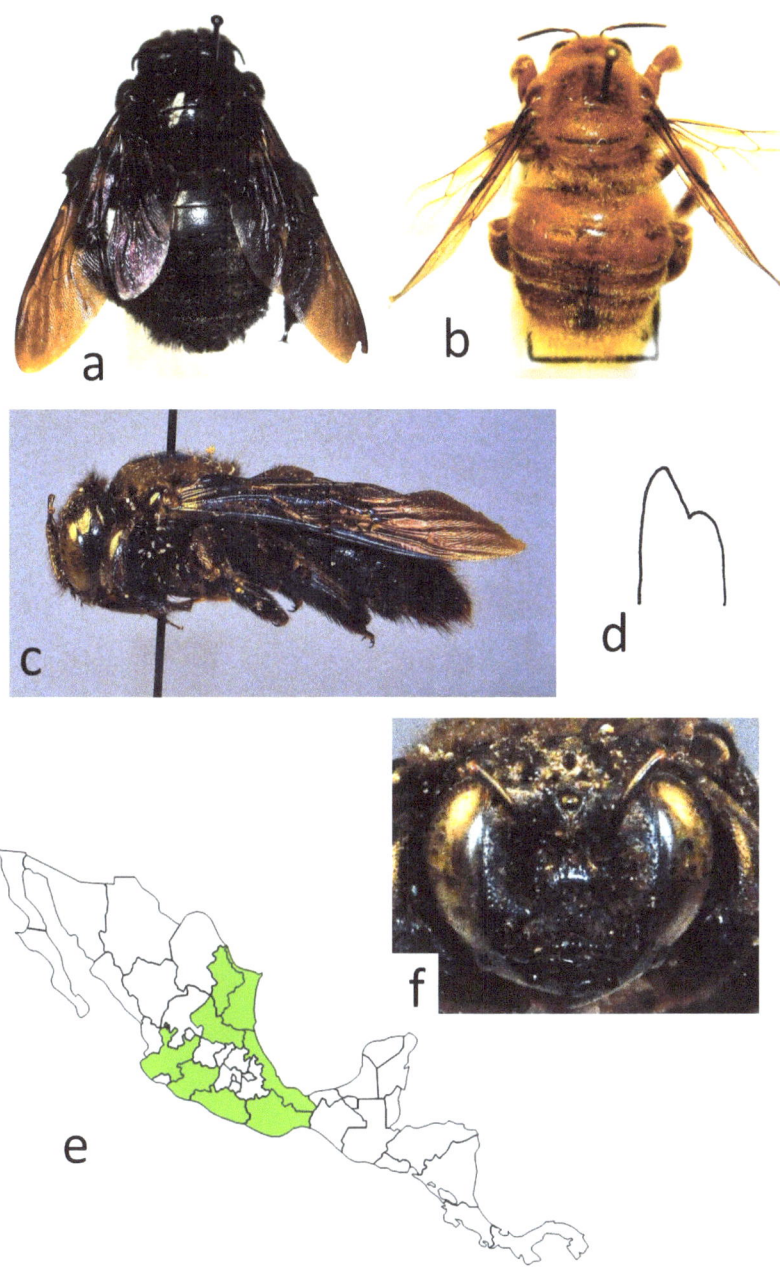

Figure 5: *Xylocopa mexicanorum*

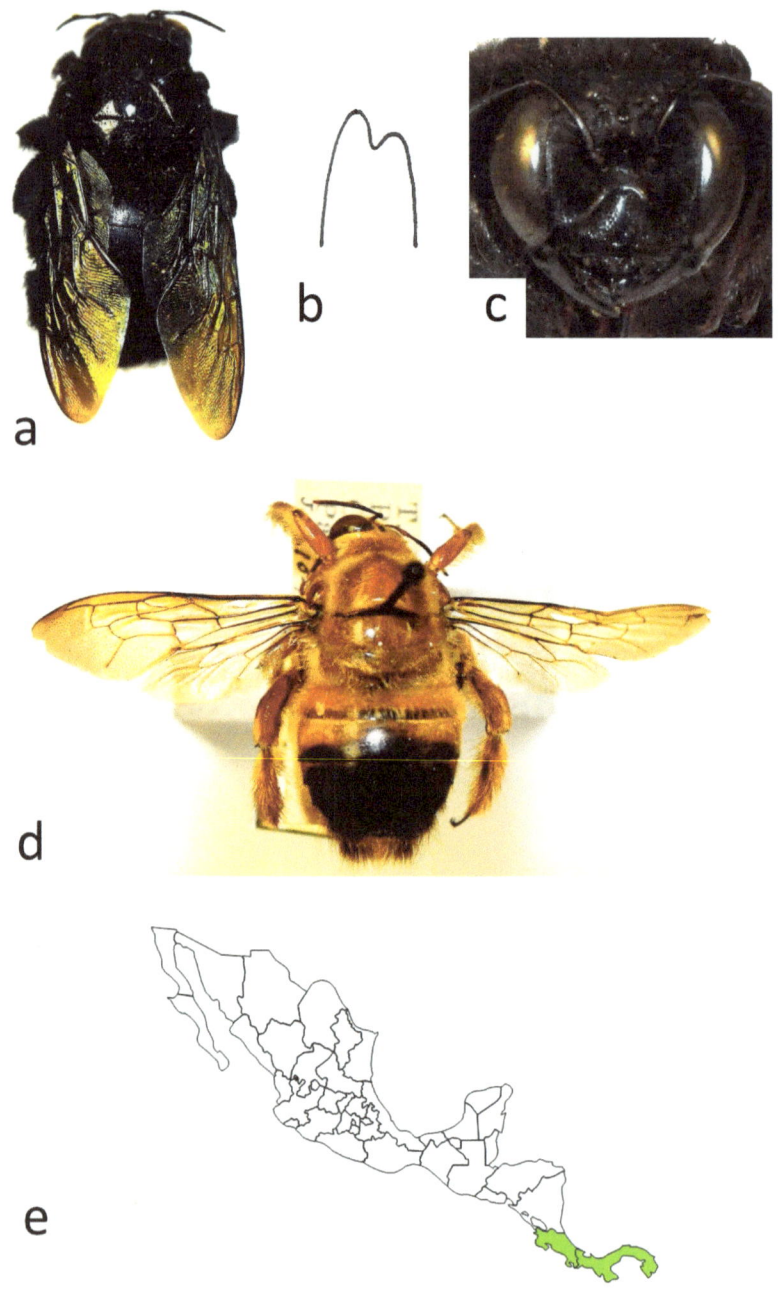

Figure 6: *Xylocopa nasica*

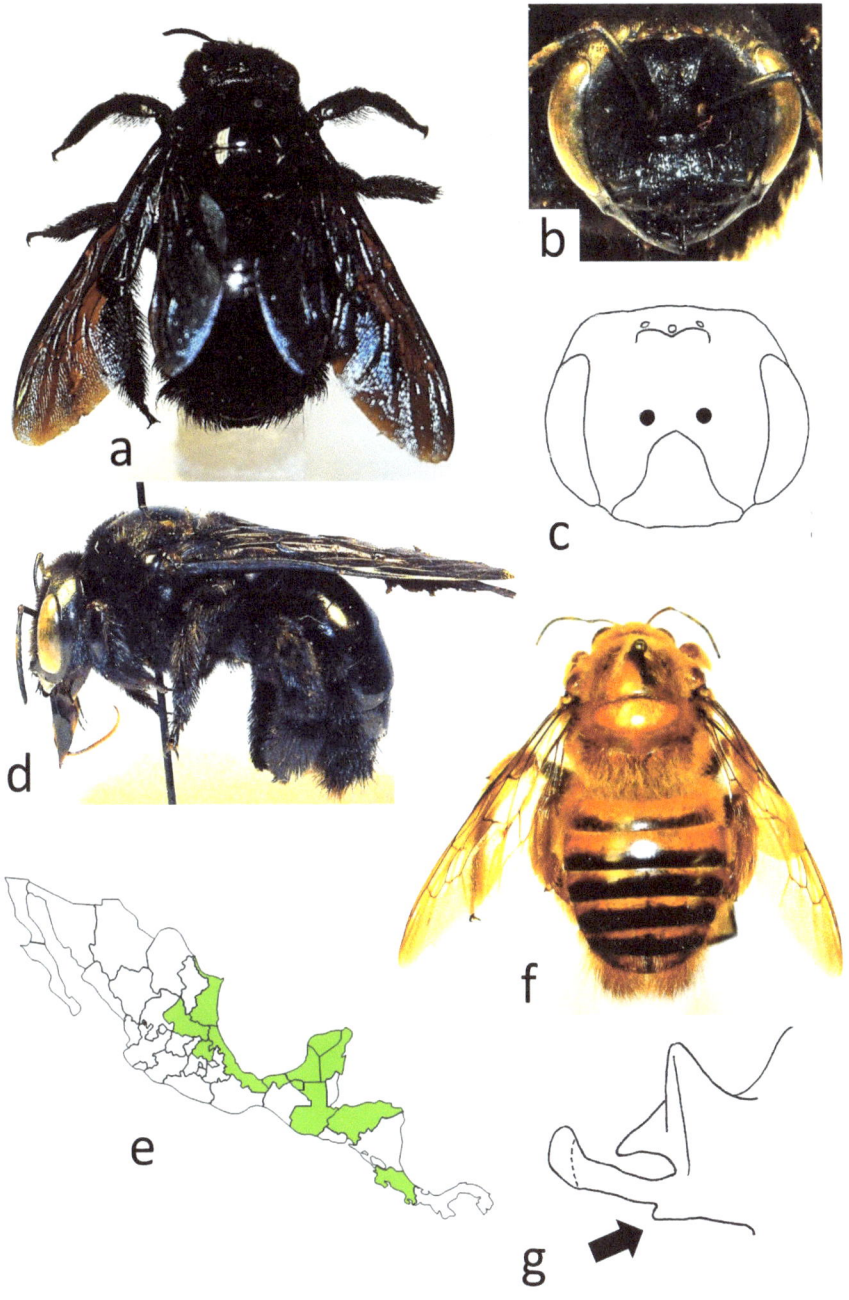

Figure 7: *Xylocopa nautlana*

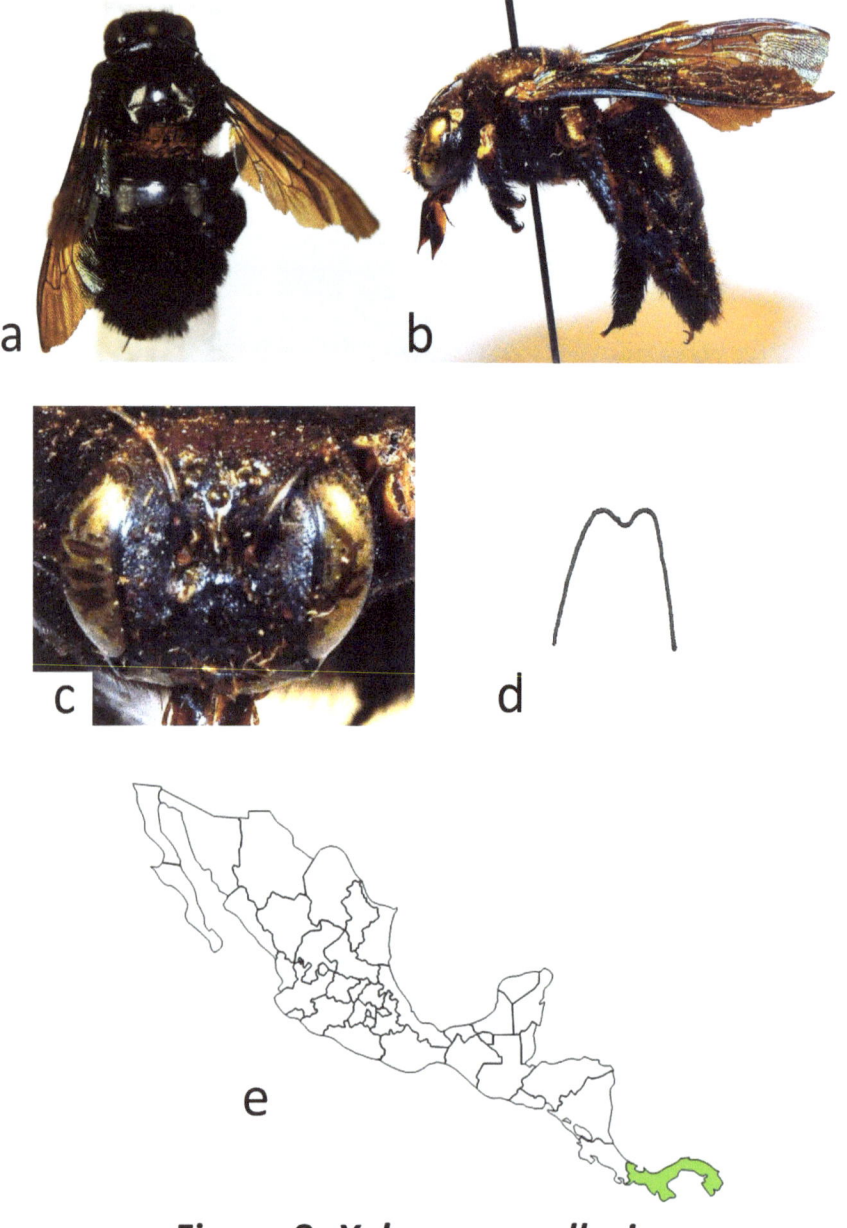

Figure 8: *Xylocopa ocellaris*

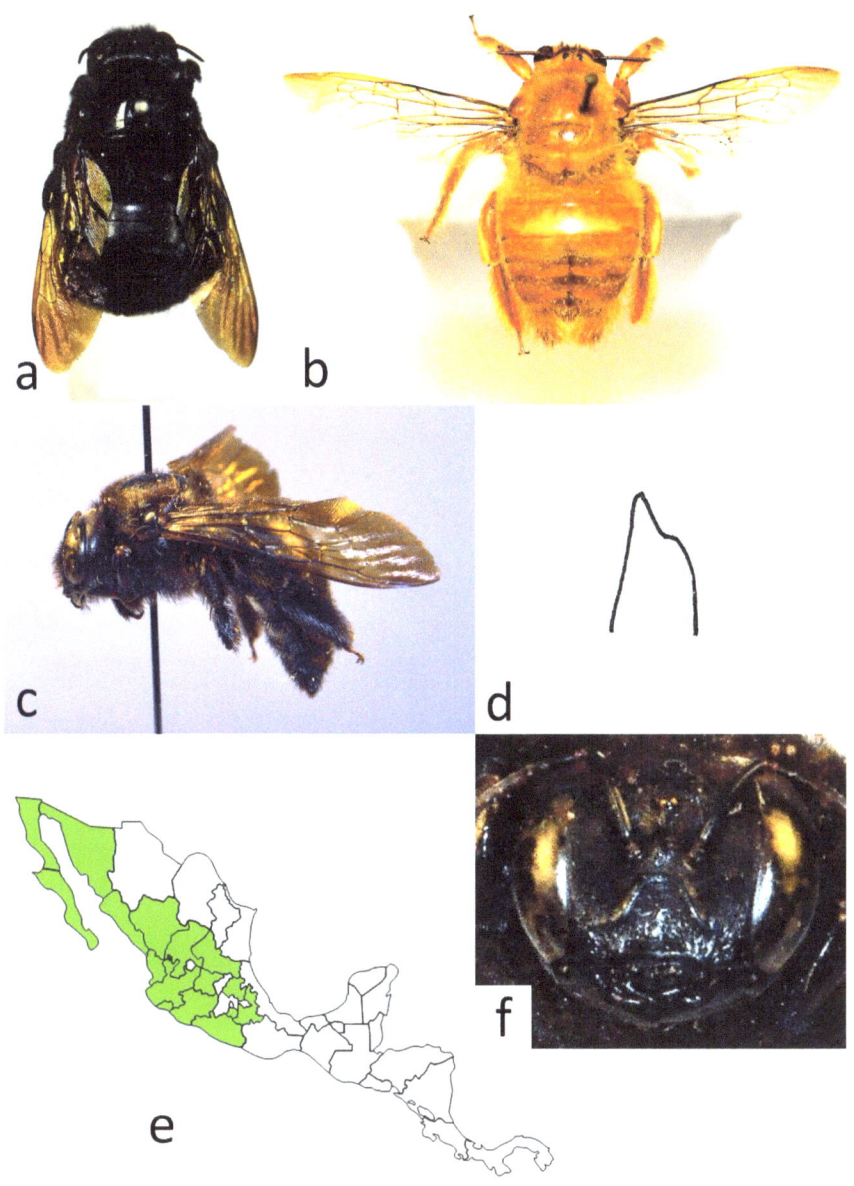

Figure 9: *Xylocopa varipuncta*

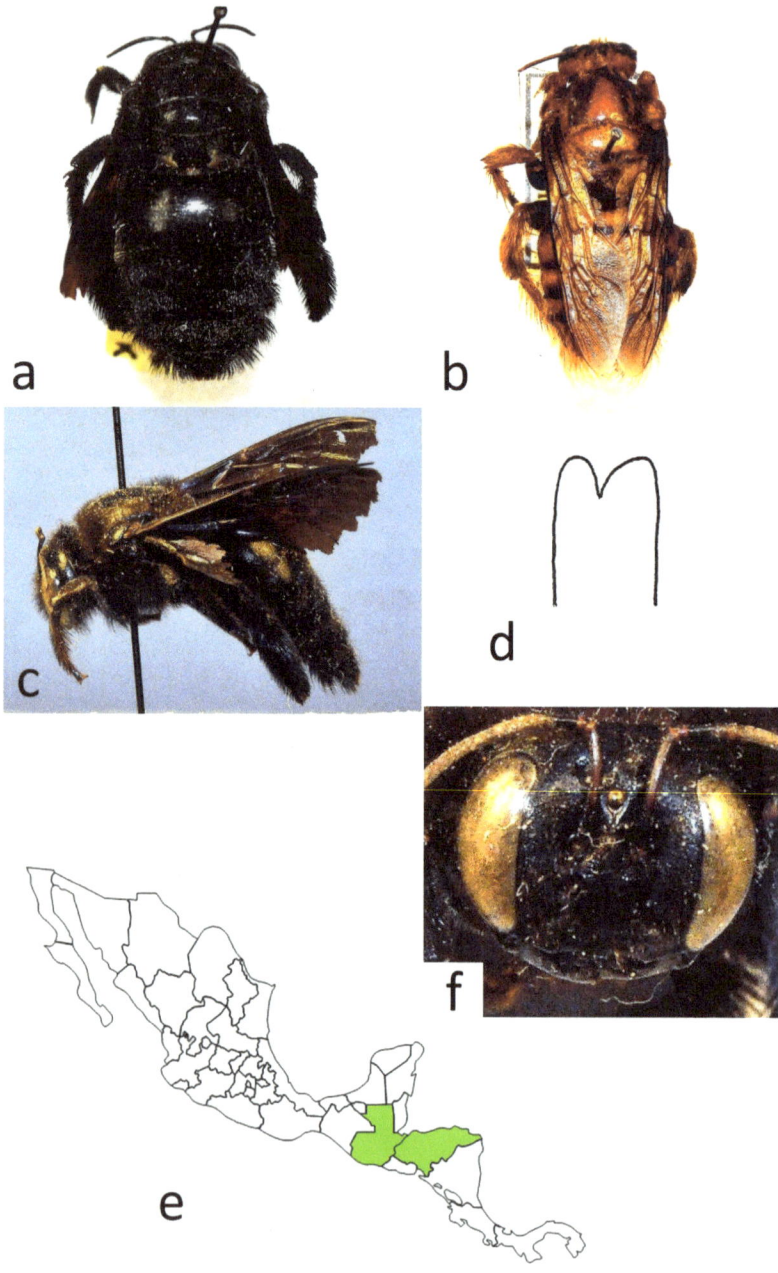

Figure 10: *Xylocopa wilmattae*

References

Cockerell, T. D. A. 1904. Descriptions and records of bees. Annals and Magazine of Natural History 7(14):21-30.

Cockerell, T. D. A. 1911. Descriptions and records of bees – XXXVIII. Annals and Magazine of Natural History 8(8):283-290.

Cockerell, T. D. A. 1912. Descriptions and records of bees – XLIV. Annals and Magazine of Natural History 8(9):554-568.

Cockerell, T. D. A. 1926. Descriptions and records of bees – CXI. Annals and Magazine of Natural History 9(17):657-665.

Cockerell, T. D. A. 1949. Bees from Central America, principally Honduras. Proceedings of the United States National Museum 98(3233):429-490.

Díaz, A. L, and Sánchez, U. S. 1998. Feeding and nesting plants of *Xylocopa cubaecola* (Hymenoptera: Apidae). Caribbean Journal of Science 34(1-2):152-155.

Dominguez-Gil, O. E. and McPheron, B. A. 1992. Arthropods associated with passion fruit in western Venezuela. Florida Entomologist 75:607–612.

Enderlein, G. 1913. Zur Kenntniss des Xylocopen Sudamerikas und uber einen Zwitter von *Xylocopa ordinaria*. Archiv für Naturgeschichte A 79(2):156-170.

Erichson, W. F. 1848. In Richard Schomburgk, Reisen in Britisch-Guiana in den Jahren 1840-1844, volume 3. Leipzig: J. J. Weber. 1260 pp.

Fabricius, J. C. 1804. Systema piezatorum. Brunswick: Reichard. xiv + 440 pp.

Gerling, D., Velthuis, H. H. W., and Hefetz, A. 1989. Bionomics of the large carpenter bees of the genus *Xylocopa*. Annual Review of Entomology 34:163-190.

Hurd, P. D. 1978. An annotated catalog of the carpenter bees (Genus *Xylocopa* Latreille) of the Western Hemisphere (Hymenoptera: Anthophoridae). Washington, D. C.: Smithsonian Institution Press. 106 pp.

Hurd, P. D. and Moure, J. S. 1963. A classification of the large carpenter bees (Xylocopini) (Hymenoptera: Apoidea). University of California Publications in Entomology 29:i-vi + 1-365.

Jackson, G. C. and Woodbury, R. O. 1976. Host plants of the carpenter bee, *Xylocopa brasilianorum* (L.), in Puerto Rico. Journal of Agriculture of the University of Puerto Rico 60:639-660.

Janzen, D. H. 1966. Notes on the behavior of the carpenter bee *Xylocopa fimbriata* in Mexico (Hymenoptera: Apoidea). Journal of the Kansas Entomological Society 39(4):633-641.

Keasar, T. 2010. Large carpenter bees as agricultural pollinators. Psyche 2010:1-7.

Latreille P. 1802. Histoire naturelle, générale et particulière, des crustacés et des insectes, tome troisiéme. Paris: L'imprimerie de F. Dufart. 467 pp.

Lepeletier, A. L. M. 1841. Histoire naturelle des insects, Hyménoptères, Suites à Buffon, volume 2. Paris: Roret. 680 pp.

Linnaeus, C. 1758. Systema naturae, Edition decima. Holmiae: Laurentii Salvii. 824 pp.

Linnaeus, C. 1767. Systema naturae, Editio duodecima reformata. Holmiae: Laurentii Salvii. 1068 pp.

Mawdsley, J. R. 2015. Cladistic analysis and evolutionary relationships of the "Megaxylocopa clade" of the genus *Xylocopa* Latreille, 1802 (Insecta: Hymenoptera: Apidae). Tropical Zoology 28(4):163-171.

Meunier, F. 1892. Observations sur quelques apides d'Ecuador. Jornal de Sciencias Mathematicas, Physicas e Naturaes (Lisboa) 2(2):63-65.

Michener, C. D. 1954. Bees of Panama. Bulletin of the American Museum of Natural History 104:1-176.

Michener, C. D. 2007. The bees of the world, second edition. Baltimore: Johns Hopkins University Press. 992 pp.

Motta Maués, M. 2002. Reproductive phenology and pollination of the Brazil nut tree (*Bertholletia excelsa* Humb. & Bonpl. Lecythidaceae) in eastern Amazonia. In: Kevan P, Imperatriz Fonseca VL, editors. Pollinating bees: The conservation link between agriculture and nature. Brasilia: Ministry of Environment. pp. 245-254.

Moure, J. S. 1960. Notes on the types of Neotropical bees described by Fabricius (Hymenoptera: Apidae). Studia Entomologica 3(1-4):97-160.

O'Farril-Nieves, H., and Medina-Gaud, S. 2007. Las plagas comunes de los árboles urbanos de Puerto Rico. Identificación y manejo. Mayagüez: International Institute of Tropical Forestry and Recinto Universitario Mayagüez. 56 pp.

Olivier, G. A. 1789. Encyclopédie Methodique, Histoire Naturelle, Tome Quatrieme, Insectes. Paris: Chez Panckouke. 331 pp.

Patton, W. H. 1879. Notes on three species of *Xylocopa*. Canadian Entomologist 11(3):60.

Pérez J. 1901. Contribution a l'etude des Xylocopes. Actes Société Linnéenne de Bordeaux 56:1-128.

Prance, G. T. 1976. The pollination and androphore structure of some Amazonian Lecythidaceae. Biotropica 8:235–241.

Smith F. 1874. Monograph of the genus *Xylocopa* Latr. Transactions of the Entomological Society of London 1874:247-302.

Waller, G. D., Vissiere, B. E., Moffett, J. O., and Martin, J. H. 1985. Comparison of carpenter bees (*Xylocopa varipuncta* Patton) (Hymenoptera: Anthophoridae) and honey bees (*Apis mellifera* L.) (Hymenoptera: Apidae) as pollinators of male-sterile cotton in cages. Journal of Economic Entomology 78:558–561.

Westwood, J. O. 1840. The natural history of bees. The Naturalist's Library 6:i-viii + 17-301 + 30 pls.

Wickler, W. 1968. Mimicry in plants and animals. New York: McGraw-Hill. 253 pp.

Glossary

Frons: The frontal area of the insect's head; in carpenter bees, the portion of the bee's head between its two compound eyes.

Integument: The hardened outer covering of the insect's body.

Mesonotum: The dorsal sclerite of the second thoracic segment in insects; in bees, the large flattened or dome-shaped sclerite on the dorsal surface of the mesosoma.

Mesosoma: The middle part of the bee's body, composed of the three thoracic segments and the first abdominal segment.

Metasoma: The hind part of the bee's body, composed of the remaining abdominal segments.

Metatibiae: The fourth segment of the hindmost (third) pair of legs.

Ocellus (plural: **Ocelli**): The simple eyes on the bee's head (as opposed to the large compound eyes).

Parameres: The lateral projections at the tip of the male genitalia in bees and other insects.

Sclerite: A hardened body part of an insect.

Tergites: Dorsal sclerites of the insect's body; in bees, usually refers to the dorsal sclerites of the metasoma.

Vestiture: The pubescence, setae, or "hairs" covering the body of an insect.

Definitions of other entomological terms can be found at the website: https://en.wikipedia.org/wiki/Glossary_of_entomology_terms

www.ingramcontent.com/pod-product-compliance
Lightning Source LLC
Chambersburg PA
CBHW041106180526
45172CB00001B/136